Collins

Mental Maths

Ages 6–7

Concepts tested

Concepts	Tests
Counting in twos	2, 3, 7, 16, 26, 29, 33, 35
Counting in tens	1–4, 6, 8, 10–12, 14, 19, 20, 26, 38
Counting in fives	5, 18, 22, 30, 31, 37, 39
Counting in threes	4, 13, 14, 20, 27, 32, 40
Counting in fours	9, 14, 16, 21, 28, 33
More / less	3, 5, 19, 25, 36, 38, 39
Number lines	1–5, 7, 11, 22, 29, 34
Number sequences	1–5, 7, 8, 12–14, 16, 18–21, 24–33, 36–40
Addition	1–18, 20–40
Subtraction	1–20, 23–40
Brackets	36, 40
Zero	4–6, 8, 9, 11–13, 15, 17, 21, 31, 33
Groups / sets	6–10, 14, 22, 33
Sharing	37–39
Place value	8, 9, 11, 12, 15–21, 26, 27, 31–35, 37, 39, 40
Fractions	1–17, 19, 22–25, 27–37
> and <	22, 24, 26, 30, 32, 35
Heavier / lighter	5, 6, 14, 18, 39
Longer / shorter / taller	15, 16, 18
Time: o'clock	1–4
Time: half past	5–7, 38–40
Time: quarter past	8, 22, 23, 25–27, 38
Time: quarter to	9, 40
Time: five past	10–16, 24
Time: ten past	17–22
Time: twenty past	28–32
Time: twenty–five past	33–38
Time: five to	40
Time: digital	2–4, 13–16, 19–21, 25–28, 30, 31, 35–40
Plane shapes	5, 6, 10, 12, 13
Solid shapes	17, 19, 20, 22–25, 38, 39
Patterns	8, 9, 12, 17, 18, 20, 37
Symmetry	25–27, 35, 40

Introduction

About this book

- This book is part of *Collins Mental Maths*, a series to support your child's development of primary mathematical skills at home.
- The grid on page 2 shows which concepts are tested in this book.
- The 40 tests provide a fun and instant way of testing your child's understanding of these concepts on a weekly basis.
- The questions are presented in a variety of simple styles to make them accessible and engaging, even to more reluctant learners.
- The tests become progressively more challenging, supporting steady advancement.
- To ensure some success every time, each test has two questions at a lower level than the rest.
- Your child may need some help with the early tests in order to get used to the format, but the consistent way in which they are presented should ensure he or she soon becomes familiar with them.
- Answers to all the questions in this book can be found on pages 45–48. There is one mark available for each question.

Recording progress

- At the end of each test is a space for you or your child to record the number of correct answers.

- 'How well did I do?' on page 4 consists of a chart on which your child can record their test marks. This not only involves your child in monitoring his or her progress but also provides practice in handling data for a purpose and involves them in the use of mathematics in an everyday situation. It also provides you with a simple visual record of how your child is meeting the demands of the tests.

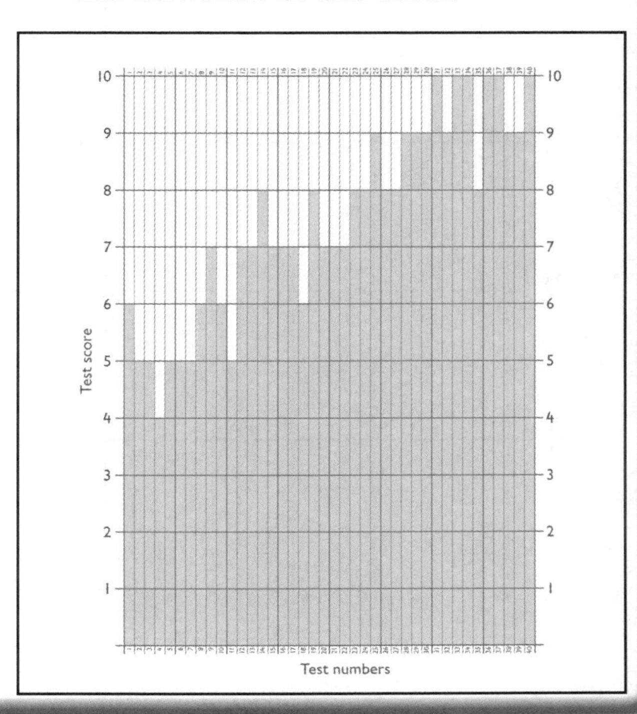

How well did I do?

Shade in your test scores on this chart.

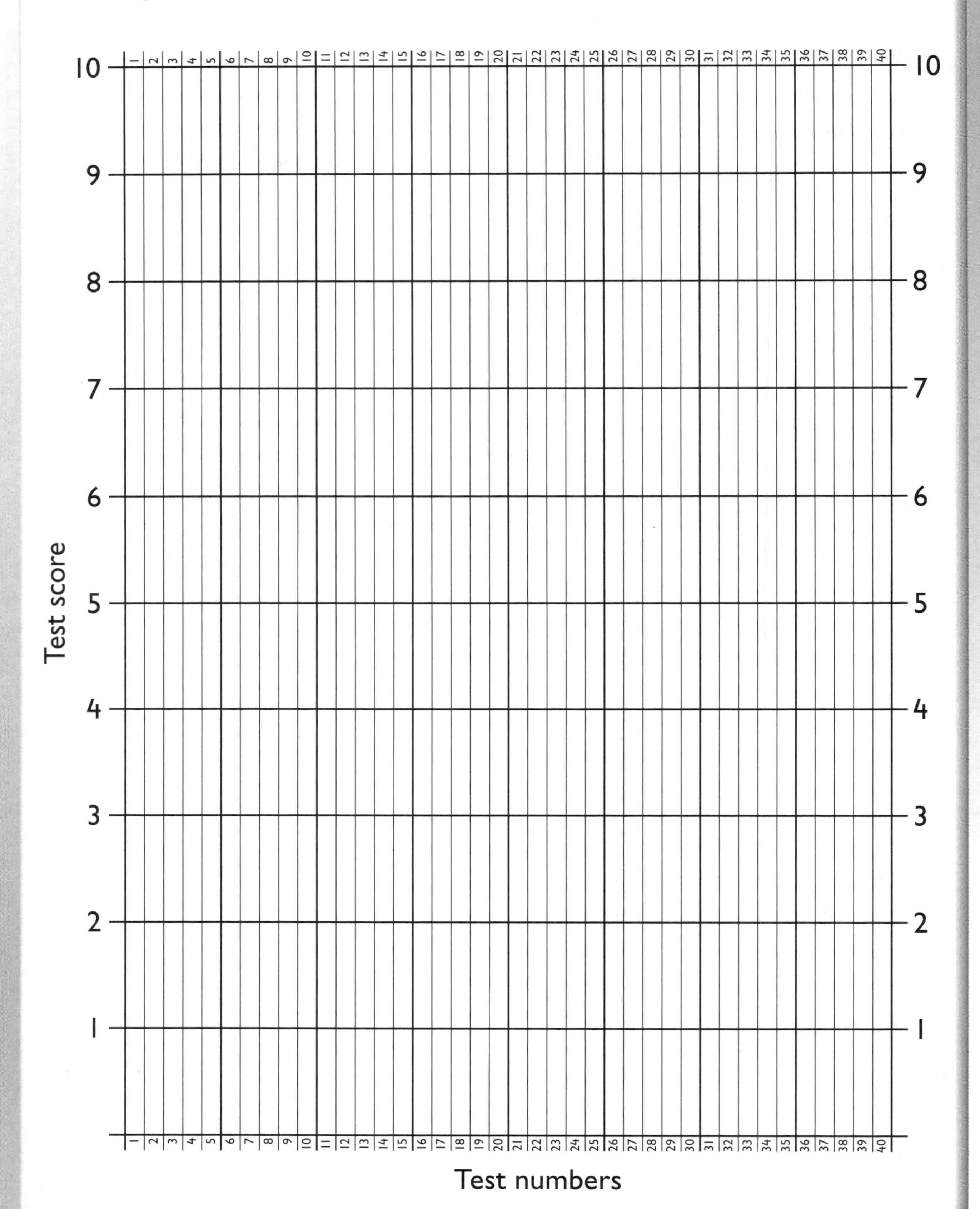

Test 1

1 1 + 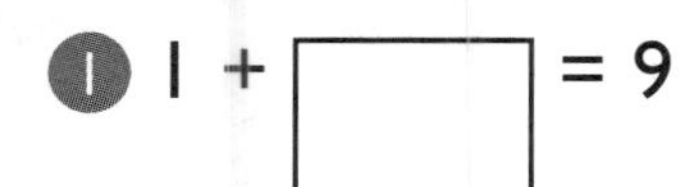 = 9

2 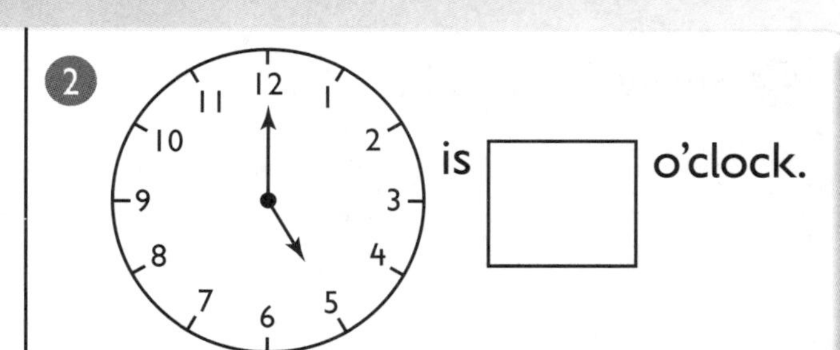

3 What number is missing?

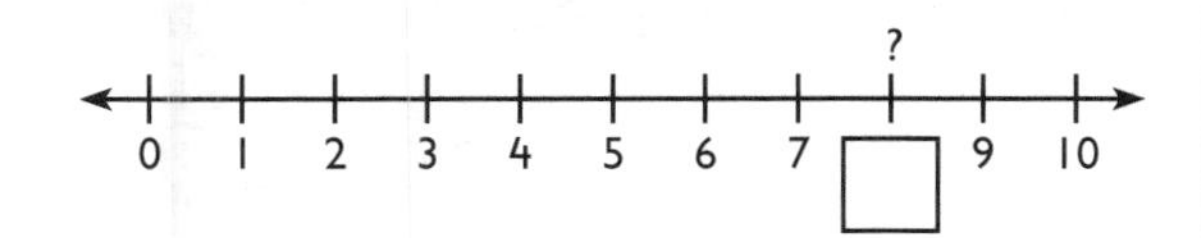

4 Which orange shows $\frac{1}{2}$ and $\frac{1}{2}$?

(a) 

(b) 

(c) 

5 Fill in the number.

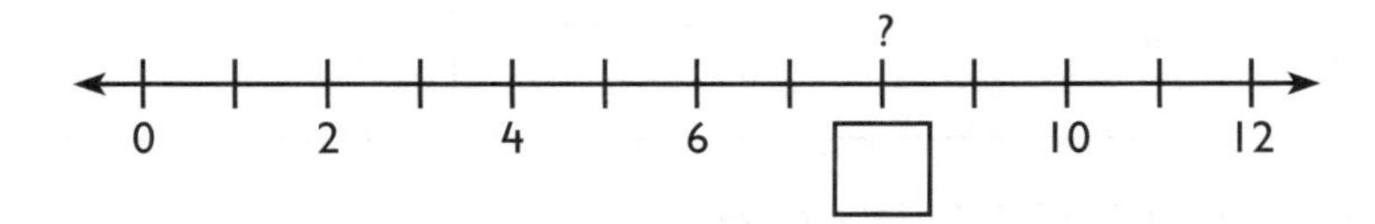

6 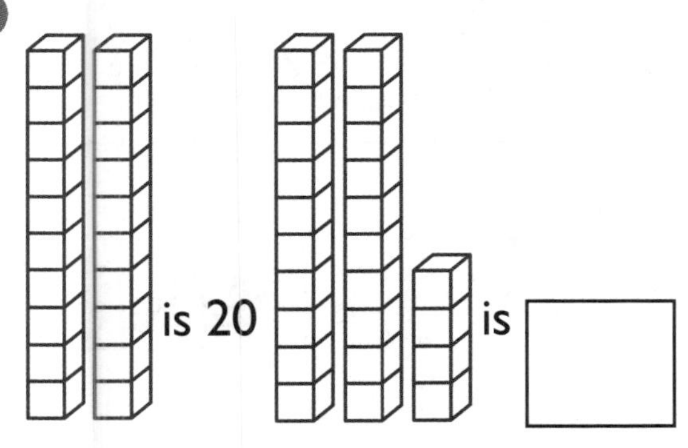

7 Which cake shows halves?

(a) 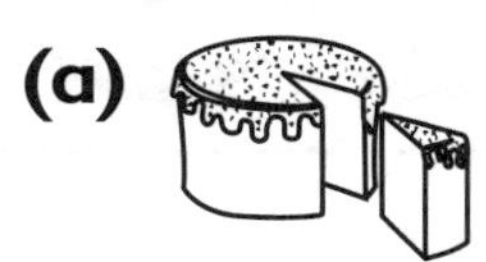

(b) 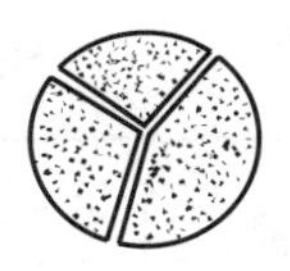

(c)

8 Fill in the missing numbers.

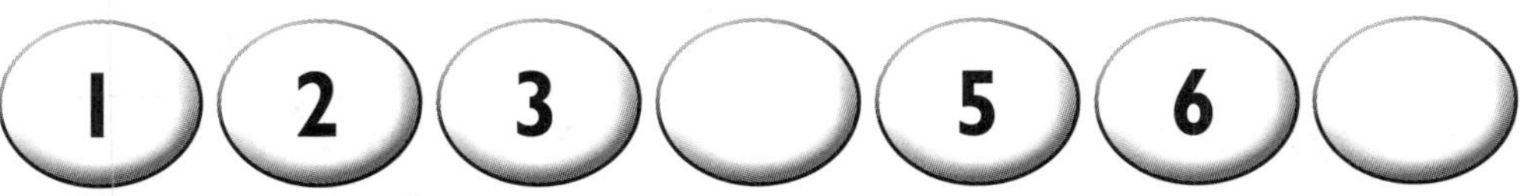

9 10 take away 2.

10 − 2 = ☐

10 8 + ☐ = 10

Score

Test 2

❶ Show 3 o'clock.

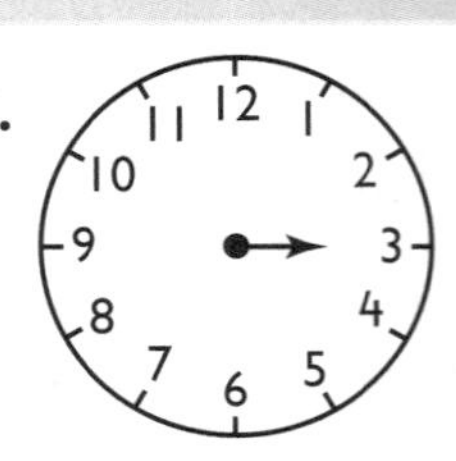

❷ Count on in 2s.
Write the numbers.

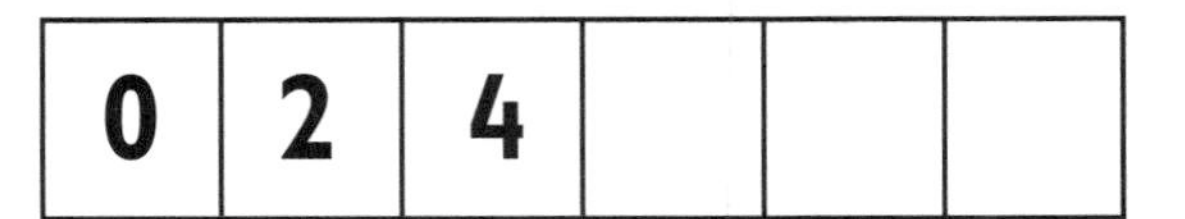

❸ Which one shows halves?

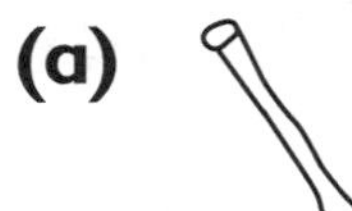

(c)

❹ 10 pins are standing.
If 2 are knocked down, how many will be left standing?

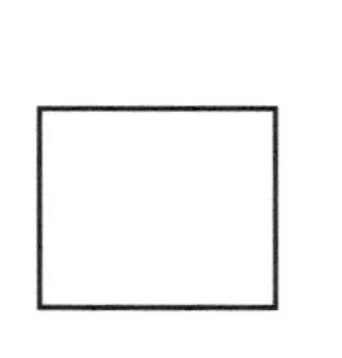
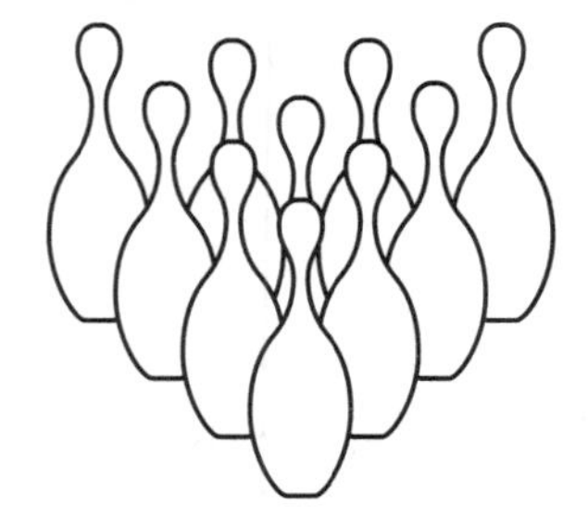

❺ Fill in the missing numbers.

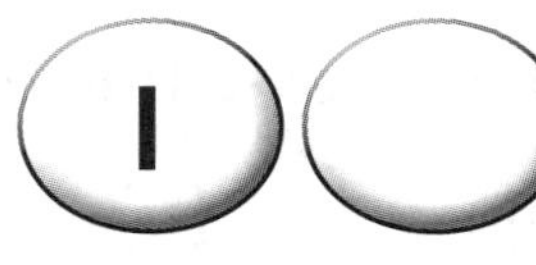

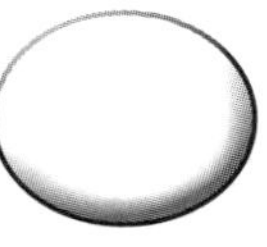

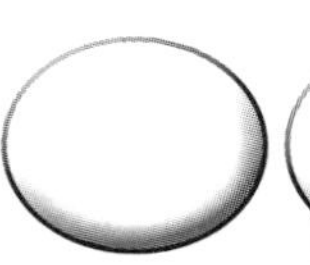

❻

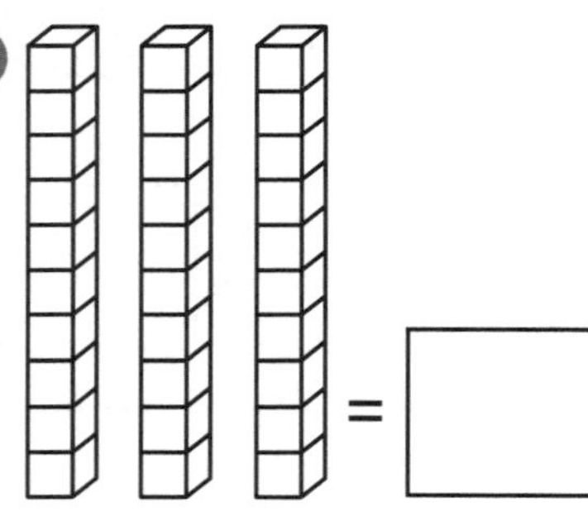

❼

is 2:00

is :

❽ 4 people have  legs.

❾ 7 + 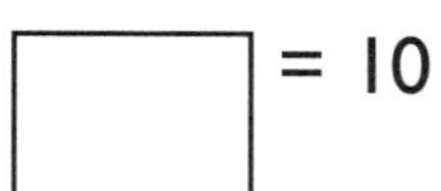 = 10

❿ 10 − 8 =

Score

Test 3

1 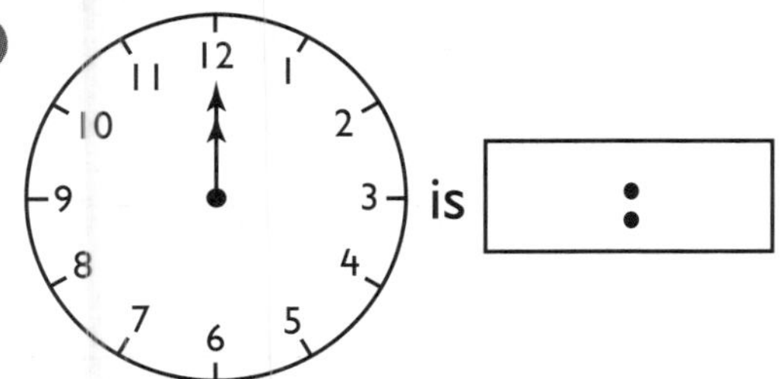 is [:]

2 2 tens = 20

3 tens = [ ]

3 Fill in the number.

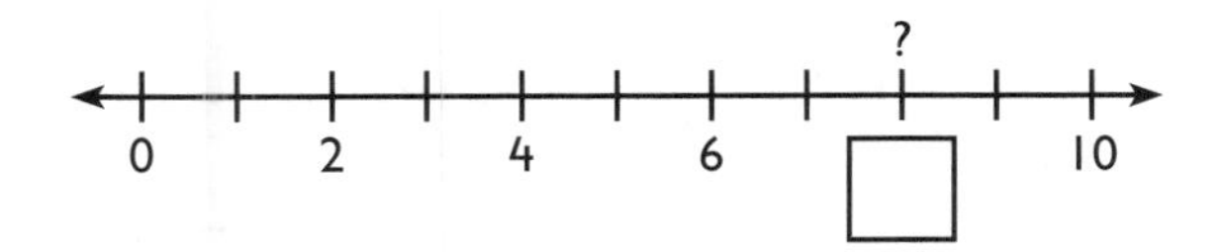

4 Ten less than 20 is 10.

Ten less than 30 is 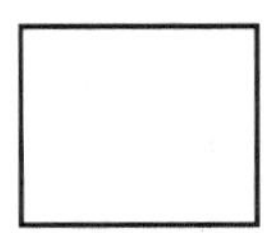

5 Count on in 2s.
Write the numbers.

10	12				20

6 Draw the hands.

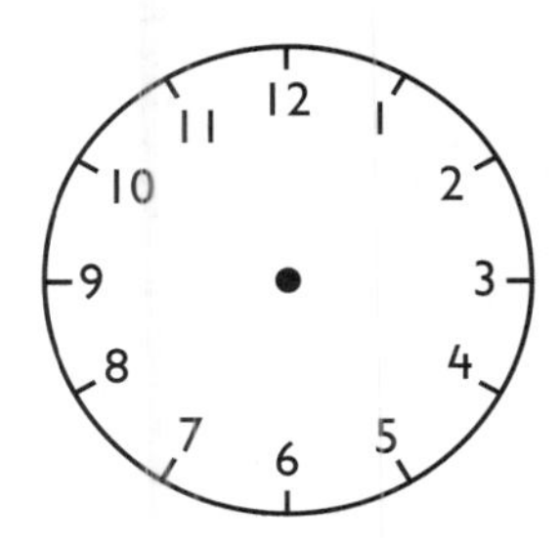

7:00

7 10 pins are standing.
If 5 are knocked down, how many will be left standing?

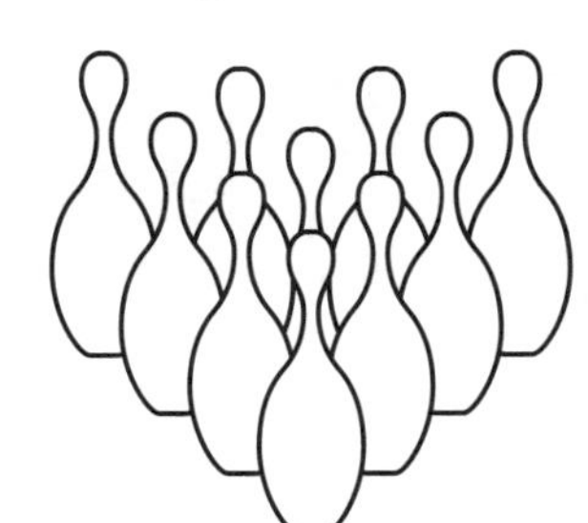

[ ]

8 [ ] + 1 = 5

9 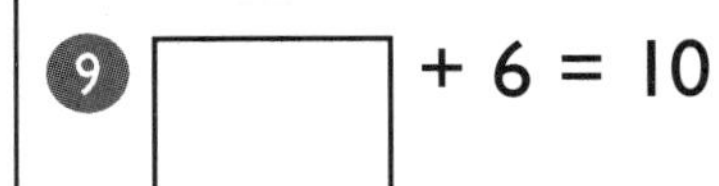 + 6 = 10

10 Which shape has $\frac{1}{2}$ shaded? 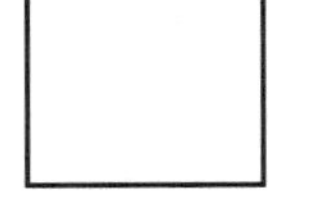

(a) 

(b) 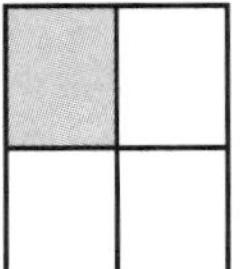

(c) 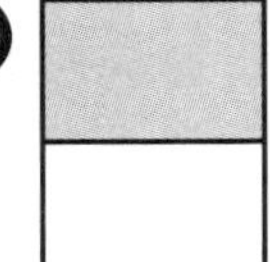

Score

Test 4

1. 3 + ☐ = 4

2. Fill in the number.

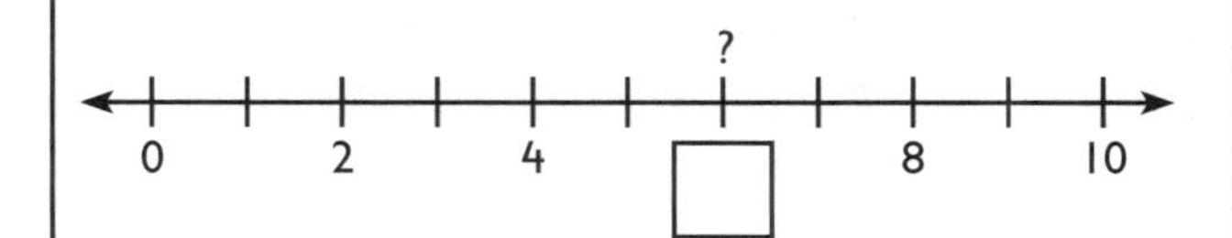

3. Shade two boxes that add up to 4.

1	2	0	5	3

4. 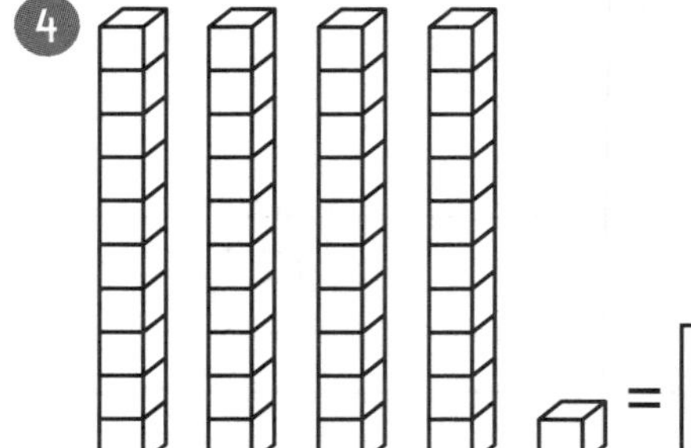 = ☐

5. Count on in 3s. Write the numbers.

3	6	9			

6. Which square is $\frac{1}{2}$ shaded?

(a) (b)

(c) 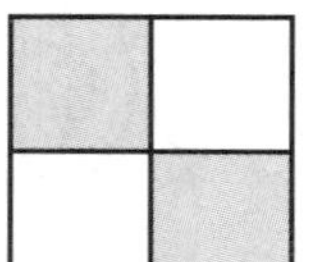

7. Show 11:00 on this clock face.

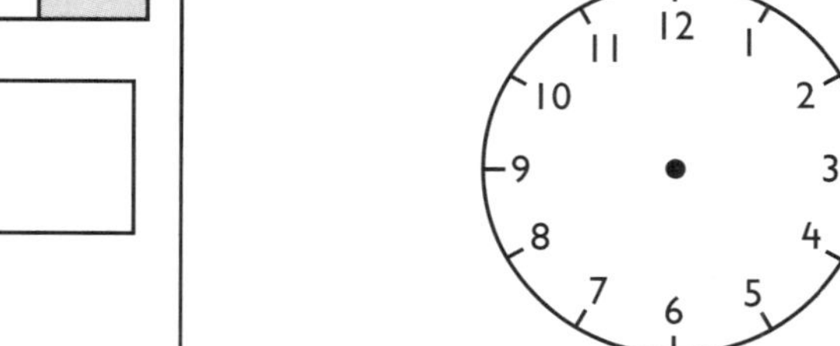

8. 3 + ☐ = 9

9. 10 − 6 = ☐

10. 10 − 5 = ☐

Score

Test 5

1. Count on in 5s.
   Write the numbers.

0	5				25

2. Fill in the number.

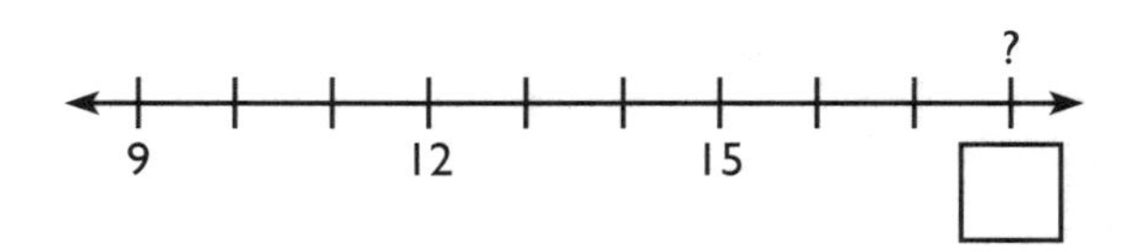

3. Who is the heavier?

4. There are ☐ squares.

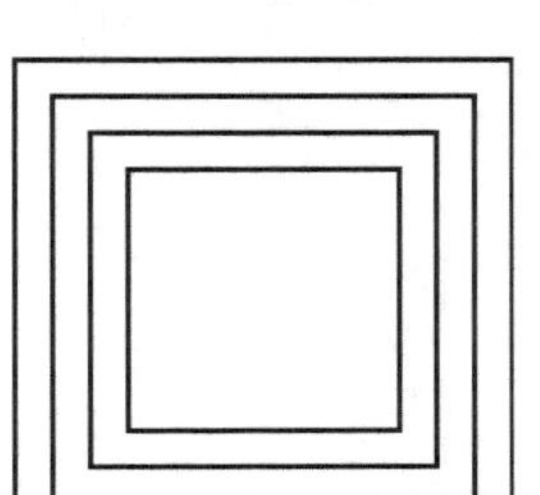

5.

half past 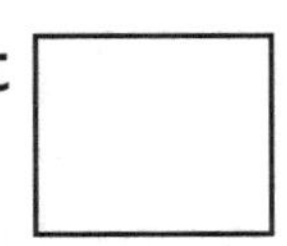

6. Which square is $\frac{1}{2}$ shaded?

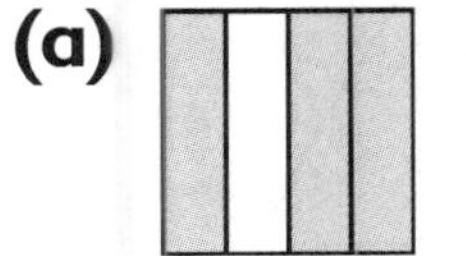

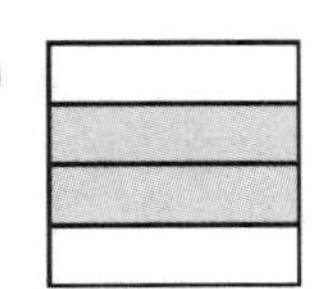

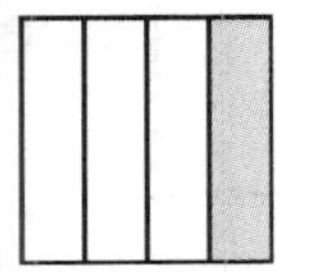

7. Which clock shows $\frac{1}{2}$ past 3?

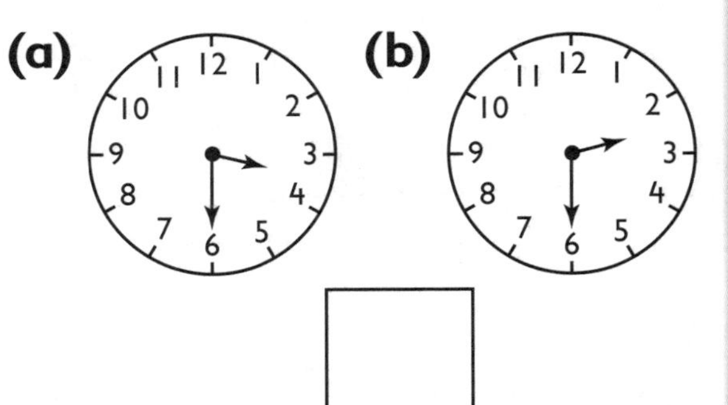

8. 8 + ☐ = 12

9. 7 − 1 = ☐

10. Ten less than 40 is ☐

Score

Test 6

1. 2 + ☐ = 10

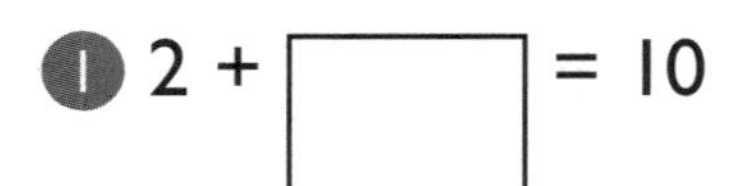

2. Write this time.

☐ past ☐

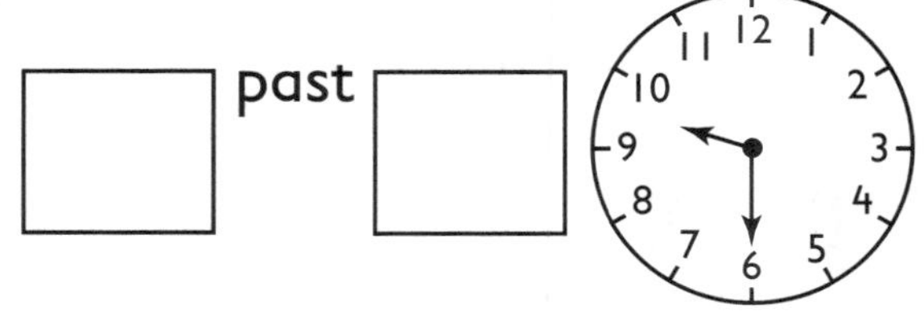

3. Which set is $\frac{1}{2}$ shaded? ☐

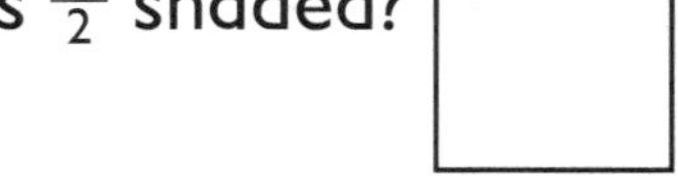

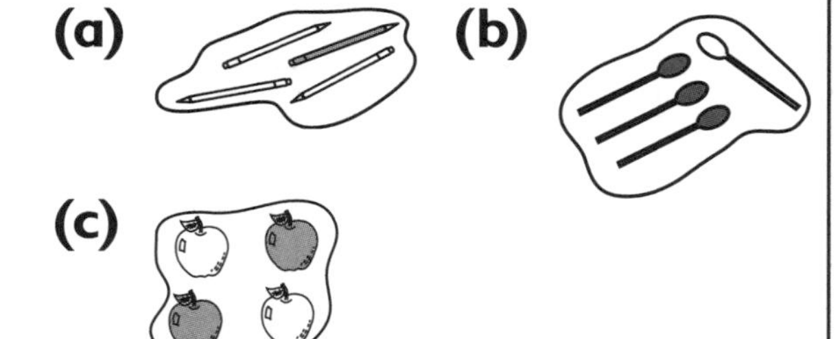

4. 10 pins are standing.
If 9 are knocked down, how many will be left standing?

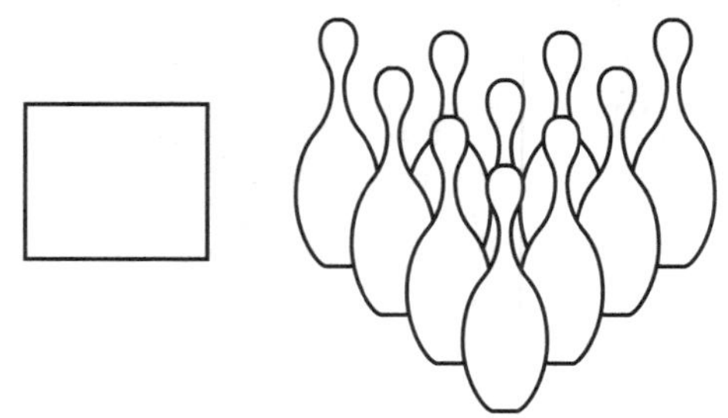

5. Fill in the missing numbers.

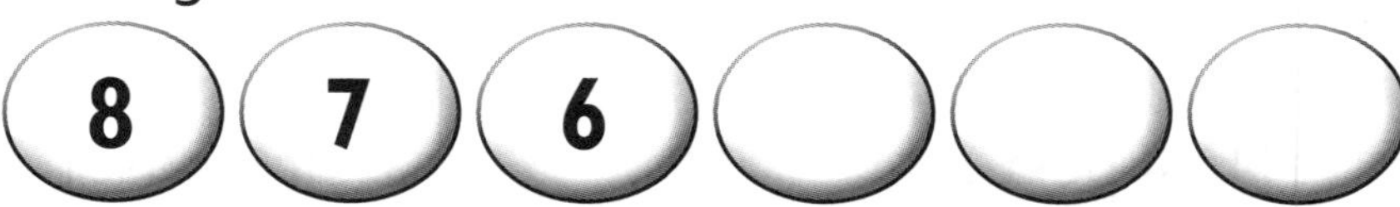

6. Who is the lighter?

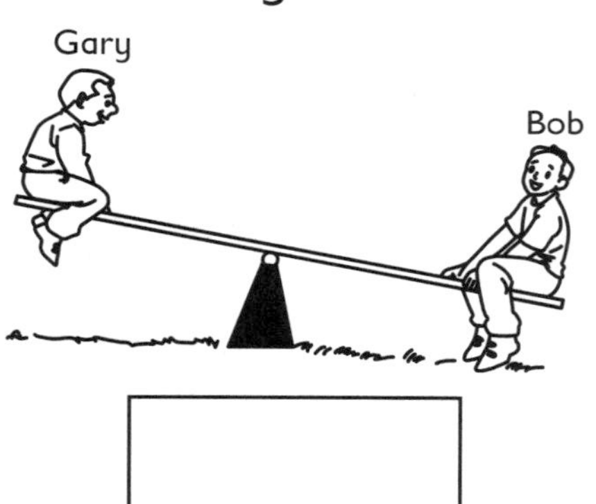

7. A triangle has 3 sides.

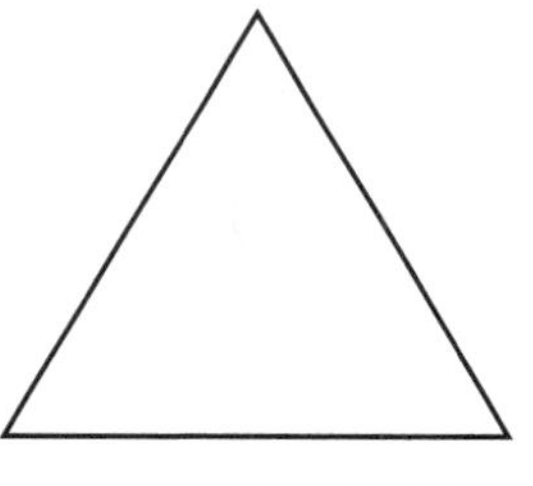

A square has ☐ sides.

8. Shade two boxes that add up to 1.

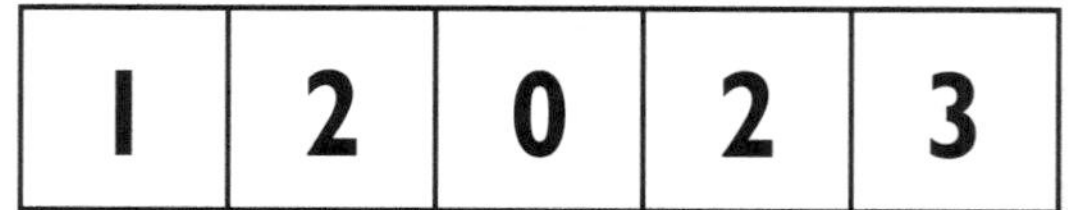

9. 5 tens = ☐

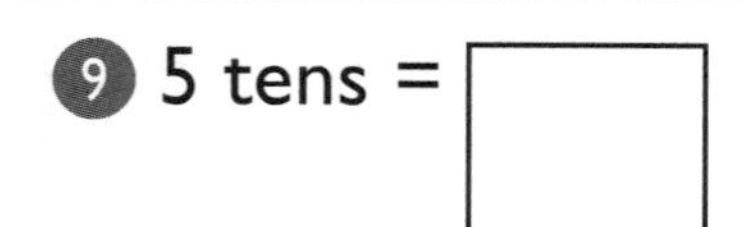

10. 9 − 2 =

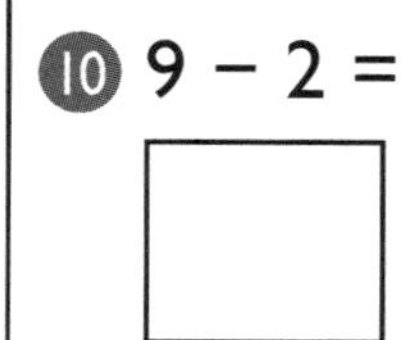

Score

Test 7

1. 4 + ☐ = 10

2. Shade $\frac{1}{2}$ of this shape.

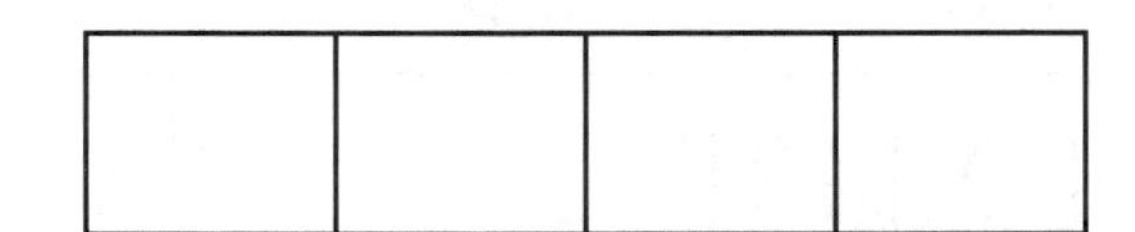

3. Fill in the number.

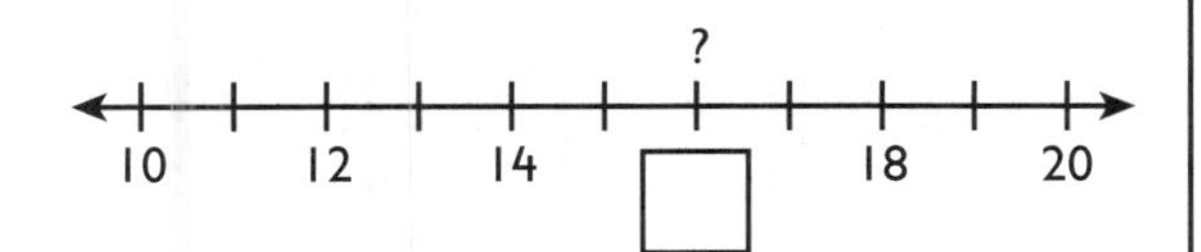

4. Draw $\frac{1}{2}$ past 11.

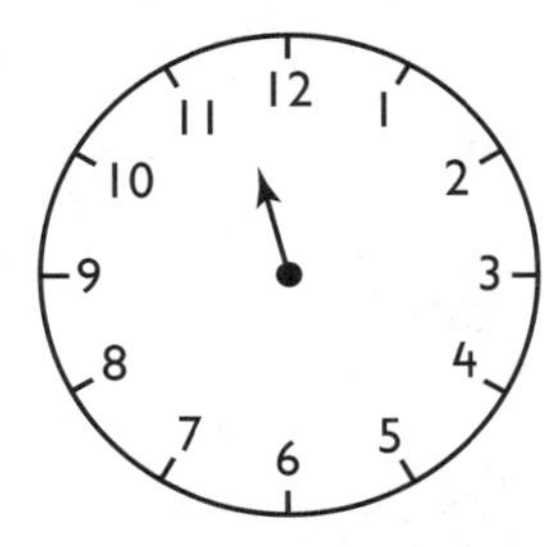

5. Fill in the missing numbers.

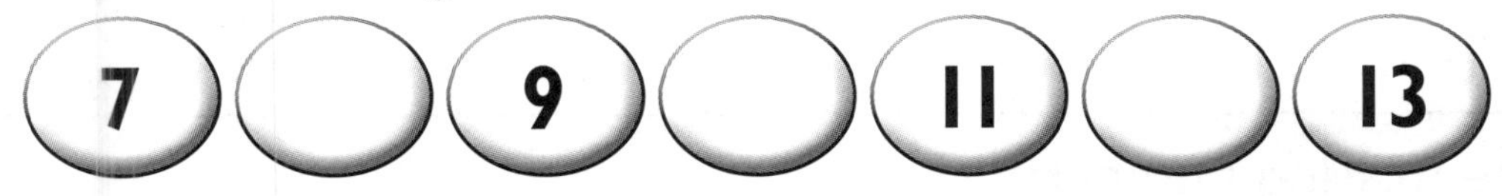

6. Shade $\frac{1}{2}$ of this set.

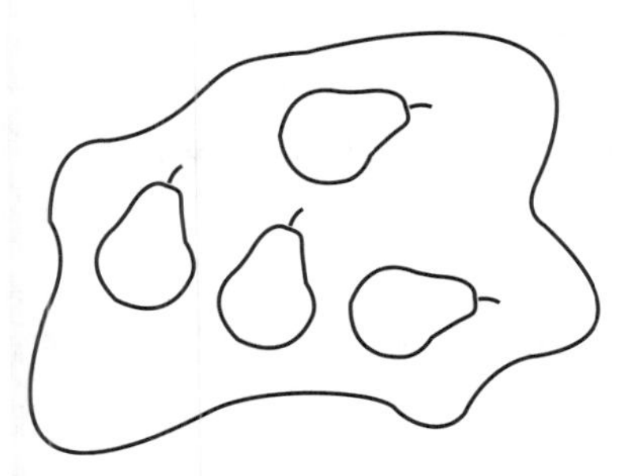

7. Count on in 2s.
Write the numbers.

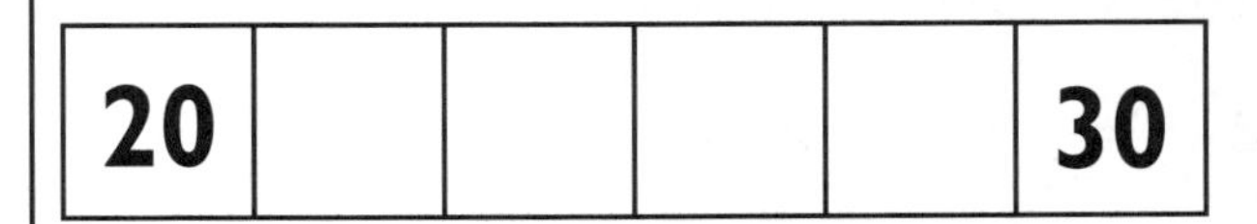

8. 5 people have ☐ legs.

9. 5 + 4 + 3 = ☐

10. 8 − 1 = ☐

Score

Test 8

1 Count on in 10s.
Write the numbers.

0	10				50

2 8 − 2 = ☐

3 Which set is $\frac{1}{2}$ shaded?

(a) (b)

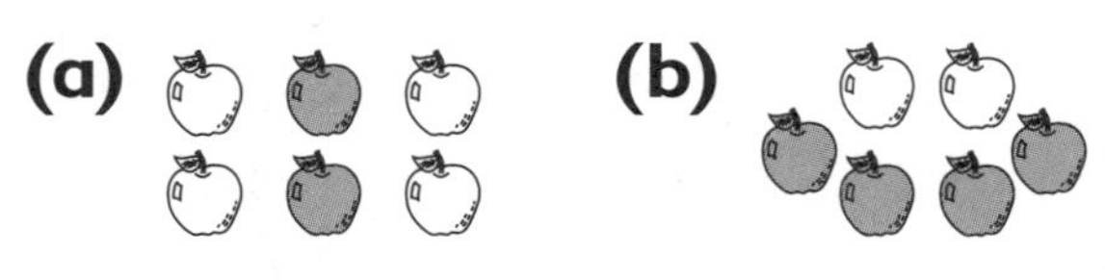

(c)

4

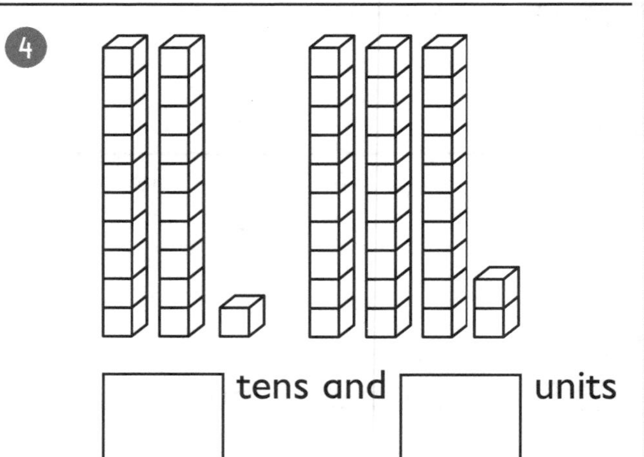

☐ tens and ☐ units

5 Fill in the missing numbers.

6 Draw $\frac{1}{4}$ past 10.

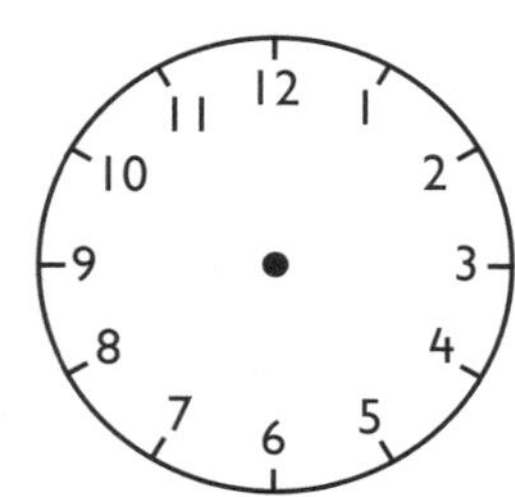

7 Shade $\frac{1}{2}$ of this set.

8 Draw the next shape.

9 ☐ − 2 = 16

10 7 + ☐ = 12

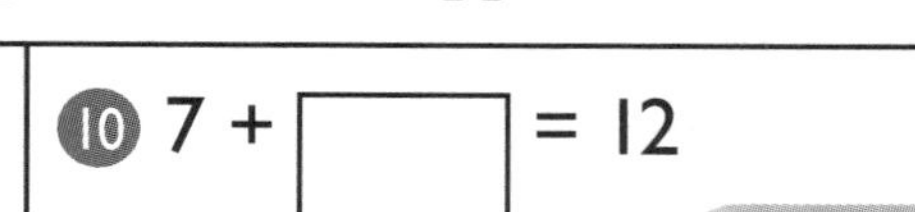

Score

Test 9

1. 40 = 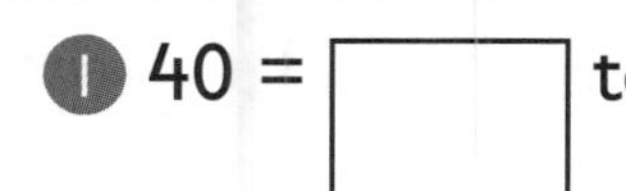 tens + 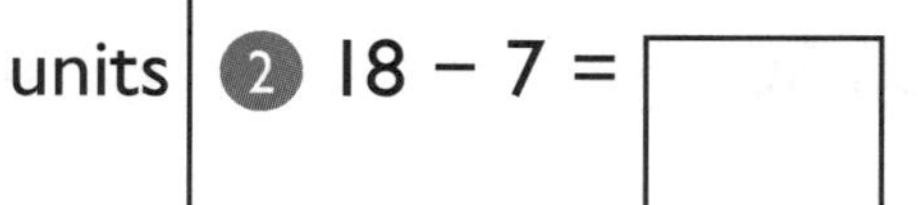 units

2. 18 − 7 = ☐

3. 10 skittles are standing.
If 6 are knocked down, how many will be left standing?

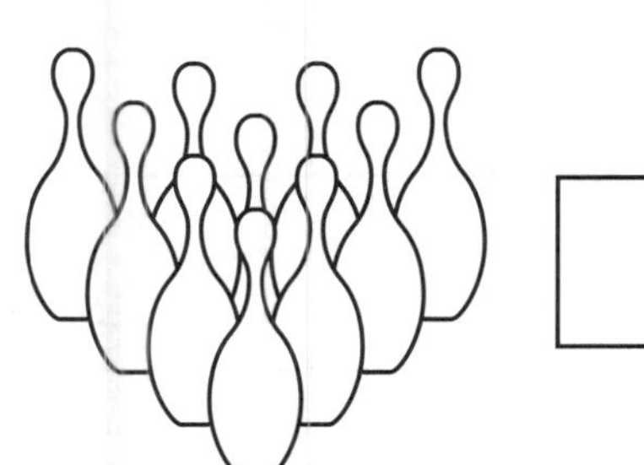

4. Shade $\frac{1}{2}$ of this shape.

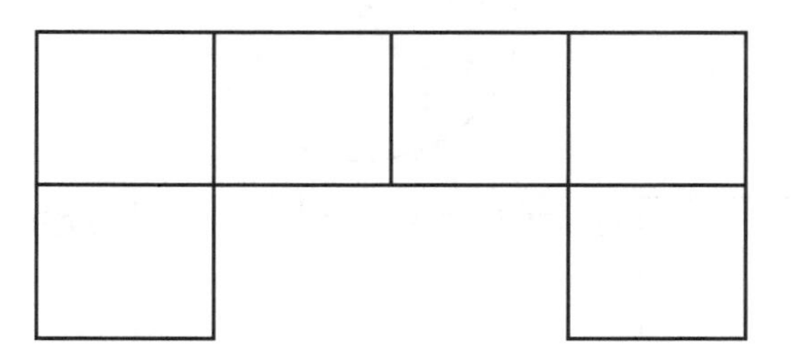

5.  Three chairs have  legs.

6. Shade two boxes that add up to 3.

1	2	3	4	3

7. 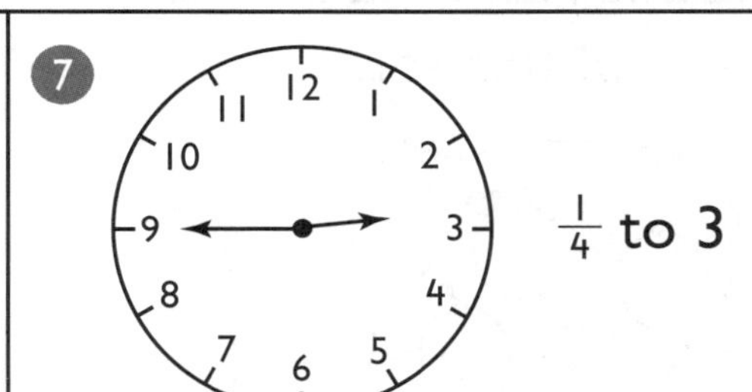

$\frac{1}{4}$ to 3

Show $\frac{1}{4}$ to 6.

8. 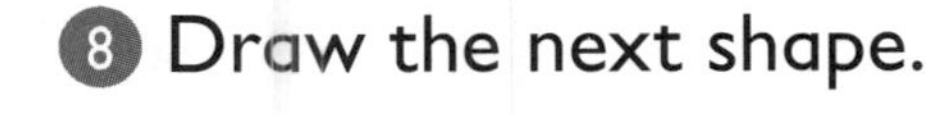 Draw the next shape.

9. 0 + ☐ = 5

10. ☐ − 1 = 22

Score

Test 10

1. 10 more than 50 is 

2. 10 less than 50 is 

3. 

☐ minutes past ☐ o'clock

4. 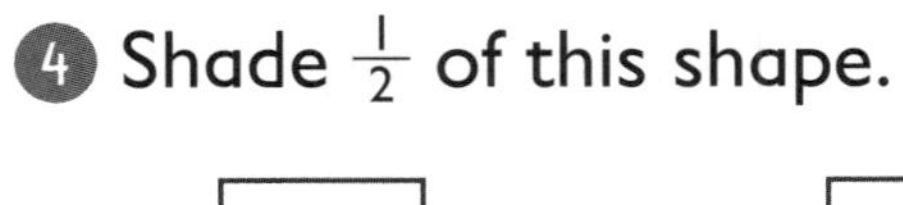

Shade $\frac{1}{2}$ of this shape.

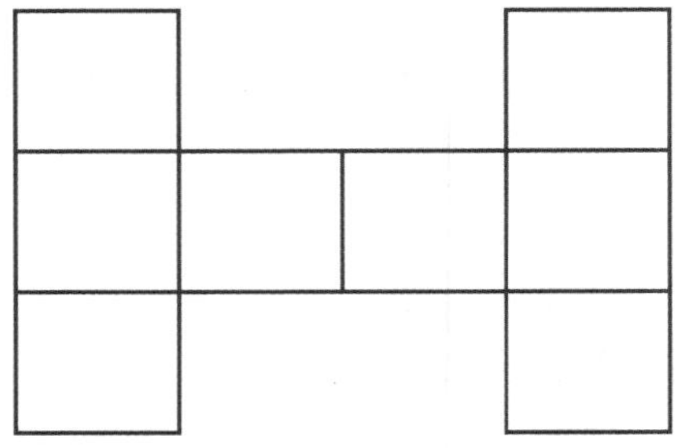

5. 

☐ minutes past ☐ o'clock

6. Shade $\frac{1}{2}$ of this set.

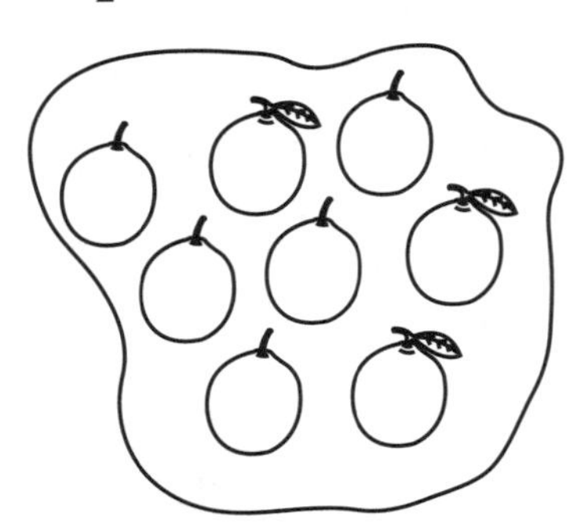

7. 10 pins are standing.
If 7 are knocked down, how many will be left standing?

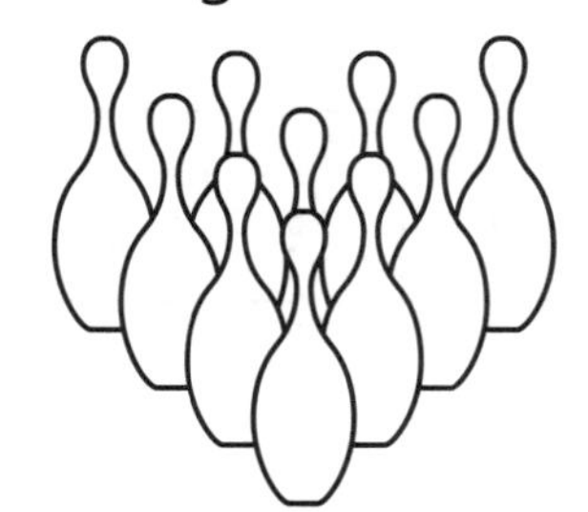

8. There are ☐ circles.

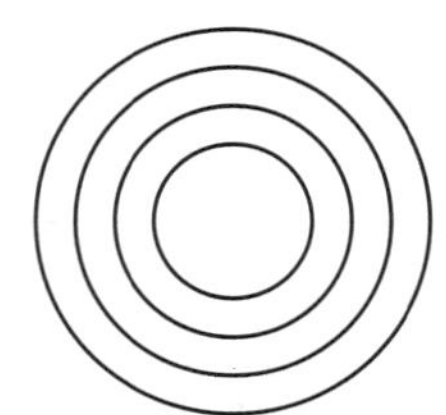

9. $12 - 5 = $ ☐

10. $10 - $ 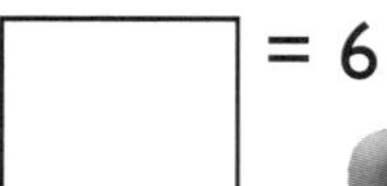 $= 6$

Score

Test 11

1. 5 +  = 7

2. Shade two boxes that add up to 2.

1	2	0	3	3

3. The time is

☐ minutes past ☐ o'clock.

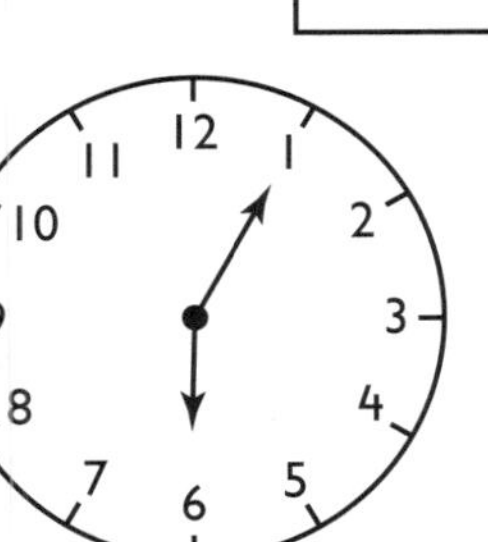

4. 10 pins are standing.
If 3 are knocked down, how many will be left standing?

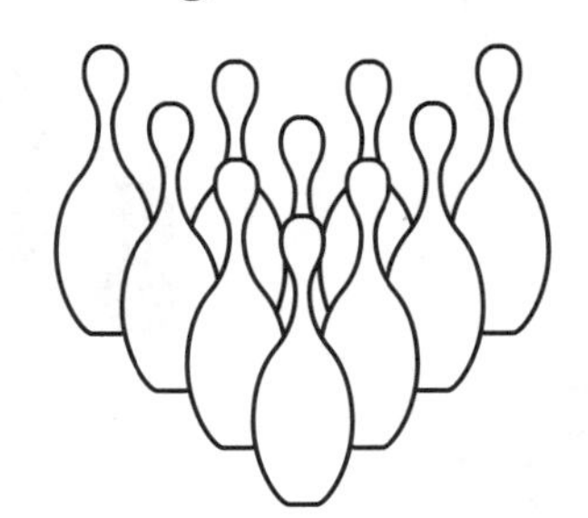

5. Fill in the numbers.

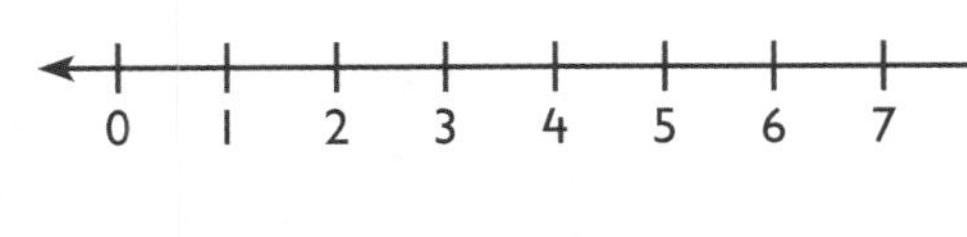

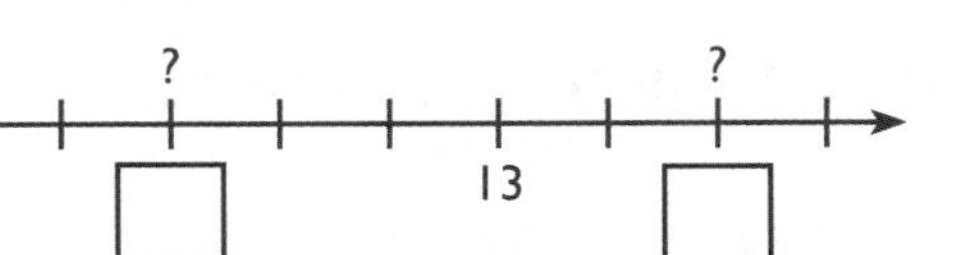

6. 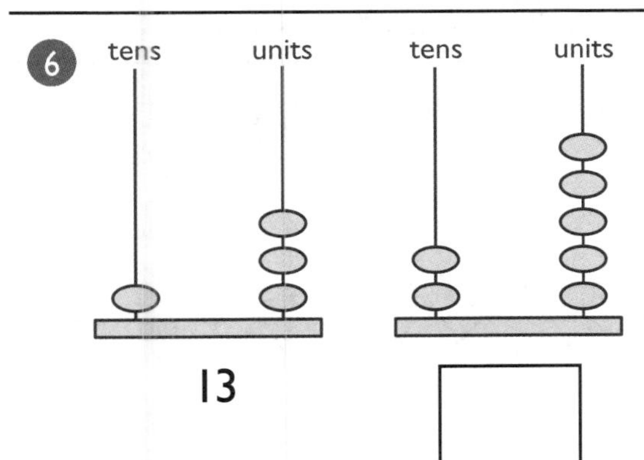

7. Shade $\frac{1}{2}$ of this shape.

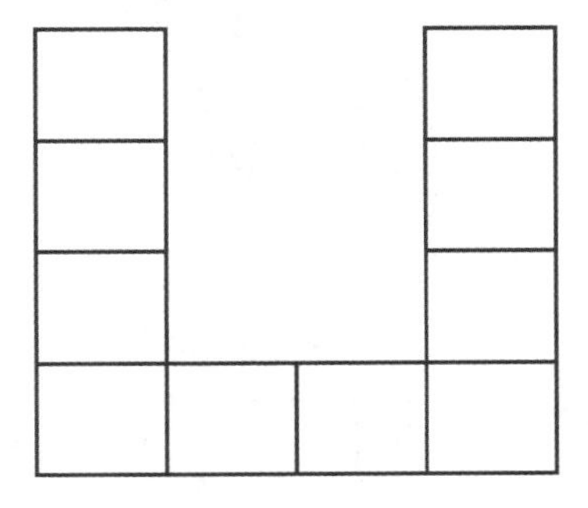

8. 5 + ☐ = 10

9. 60 − 10 = ☐

10. $\frac{1}{2}$ of 8 = ☐

Score

Test 12

1 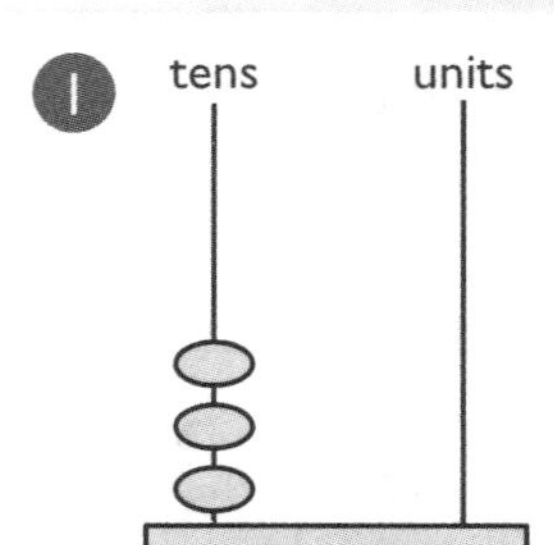

= 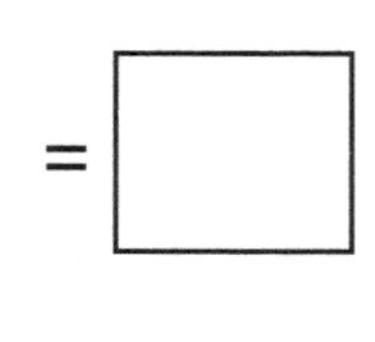

2 Shade two boxes that add up to 5.

1	2	0	1	3

3 Which set is $\frac{1}{2}$ shaded?

(a) 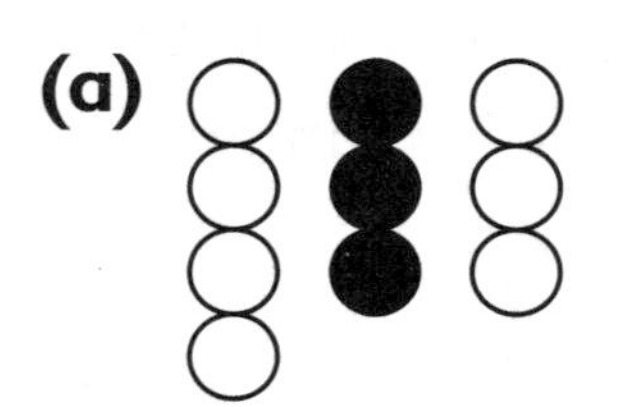

(b) 

(c) 

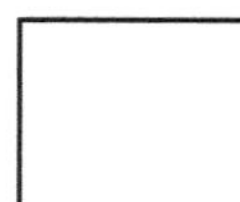

4 There are ☐ triangles.

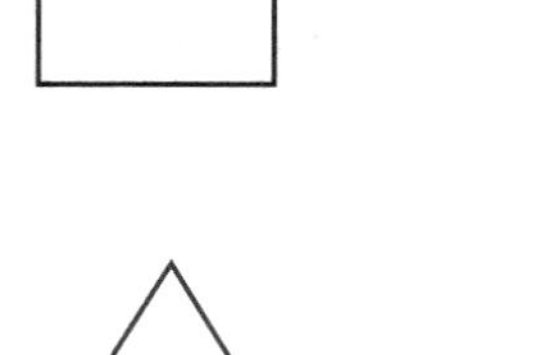

5 Fill in the missing numbers.

 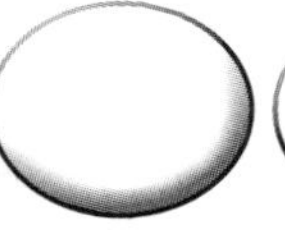 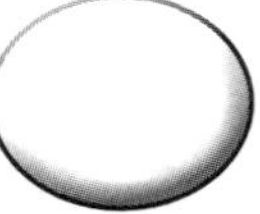 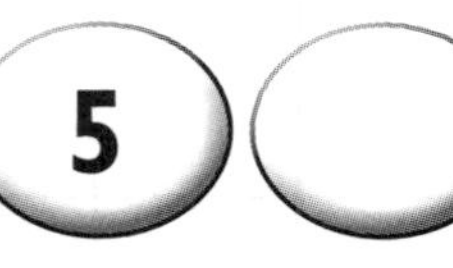

10 | 9 | 8 | ☐ | ☐ | 5 | ☐

6 

The time is ☐ past ☐

7 Draw the next shape.

8 4 + 4 + ☐ = 12

9 10 less than 70 is ☐

10 9 − 1 = ☐

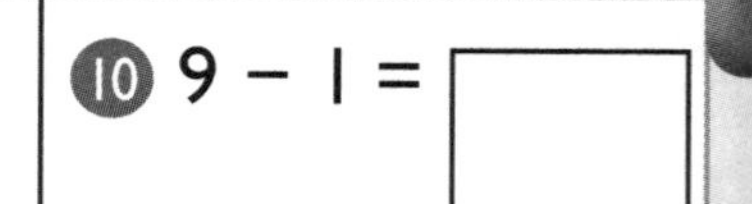

Score

Test 13

1. $\frac{1}{2}$ of 10 =

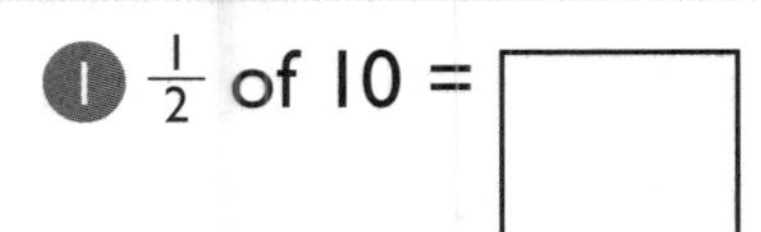

2.

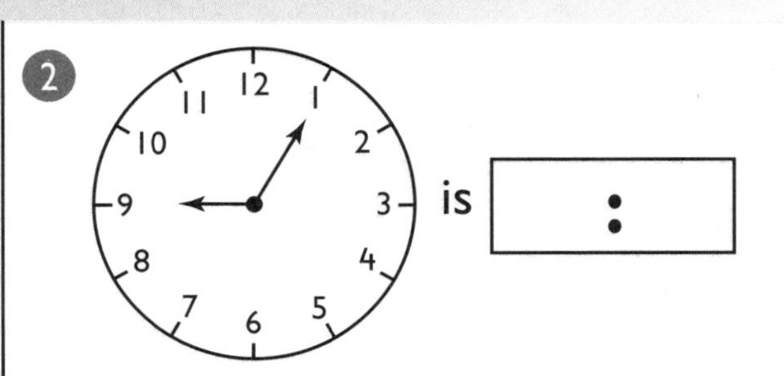

3. Shade the piece that fits in the space.

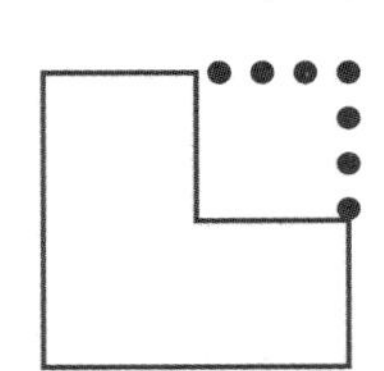

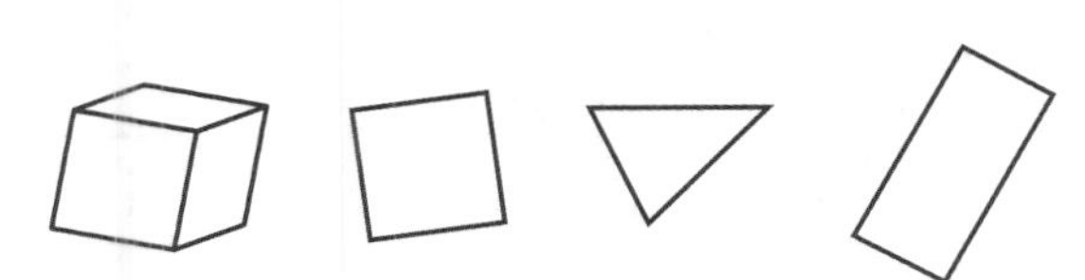

4. Saqib ate 4 chocolates and had 9 left.
How many did he have at the start?

5. Shade three boxes that add up to 3.

6. Shade $\frac{1}{2}$ of this shape.

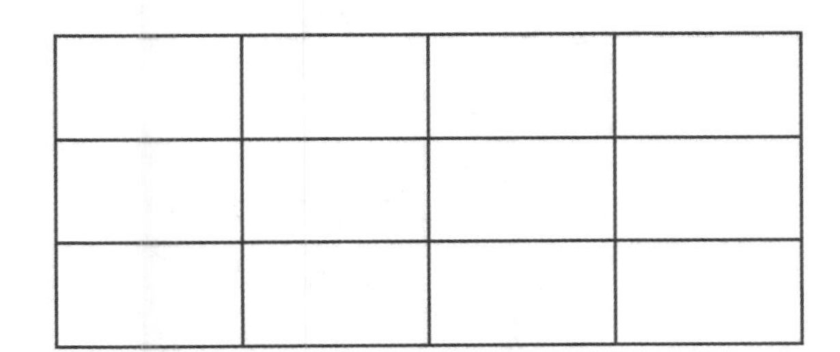

7. Count on in 3s.
Write the numbers.

0	3				15

8. 5 minutes past 10 o'clock is

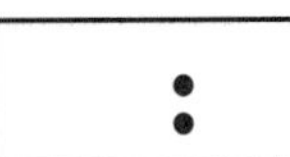

9. ☐ − 3 = 9

10. 5 − 3 =

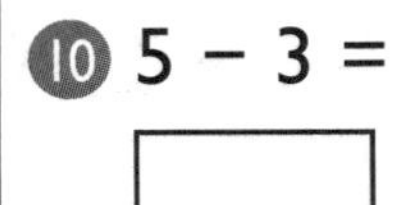

Score

Test 14

1 

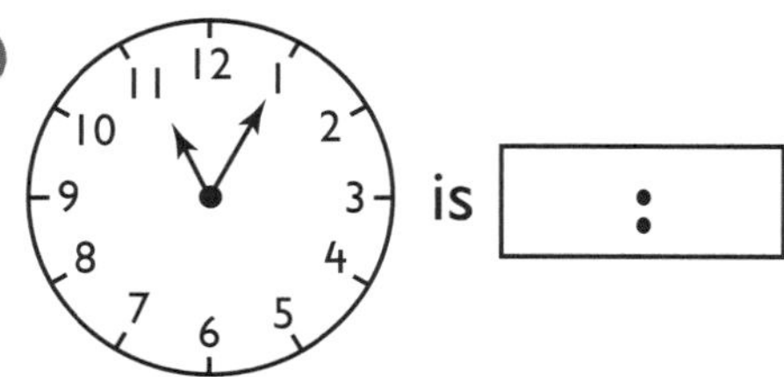

2 Count on in 3s.
Write the numbers.

12	15			24		

3 Which set has $\frac{1}{2}$ shaded?

(a) 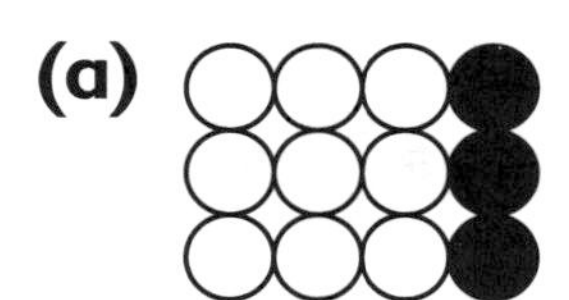

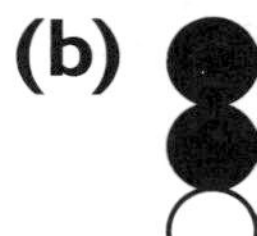

(b) 

(c) 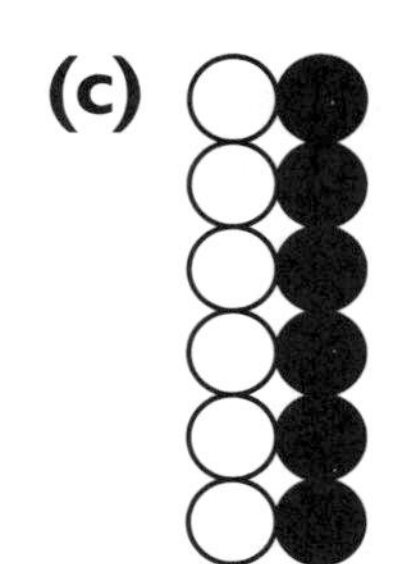

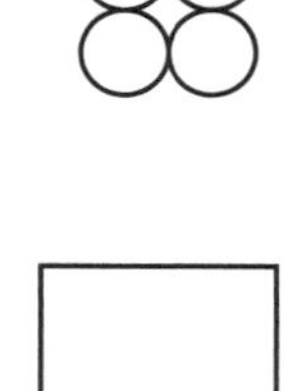

4 10 pins are standing.
If 8 are knocked down, how many will be left standing?

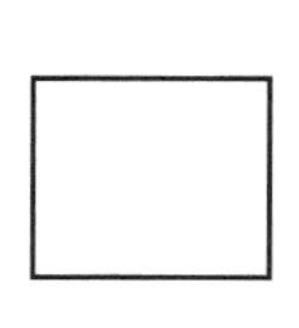

5 

Four stools have  legs.

6 Which is heavier, A or B?

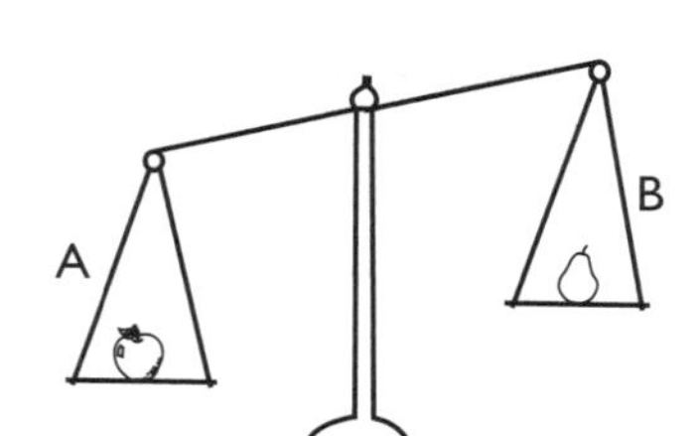

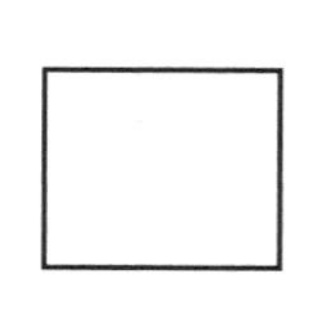

7 Shade $\frac{1}{2}$ of this set.

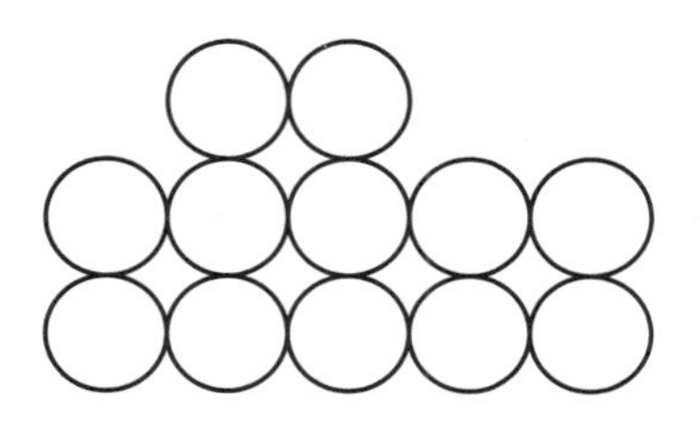

8 Fill in the missing numbers.

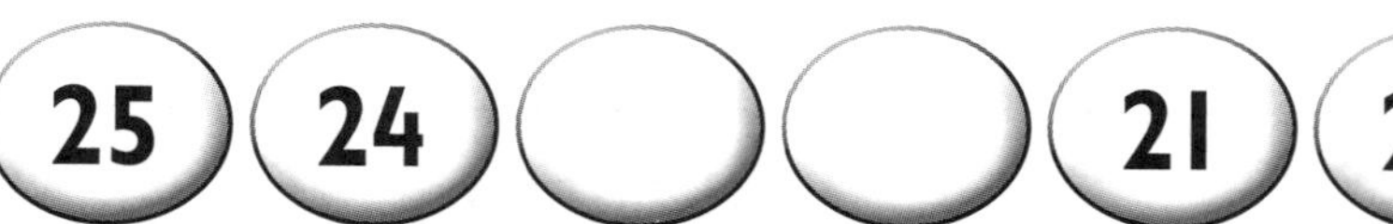

9 10 less than 80 = 

10  − 5 = 5

Score

Test 15

1. 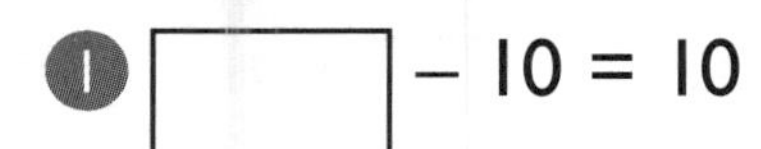

   $\square - 10 = 10$

2. Show 50 on this abacus.

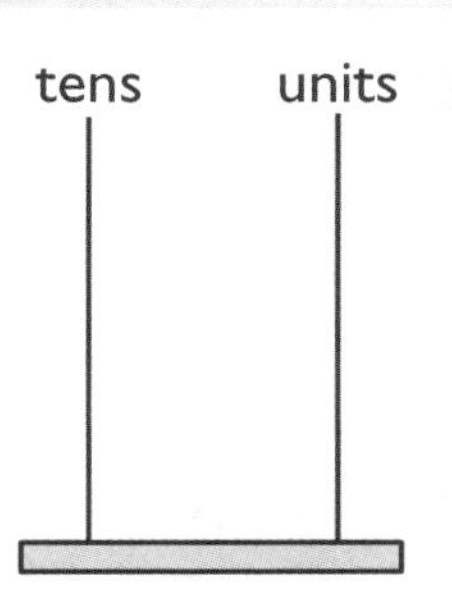

3. Show 6:05

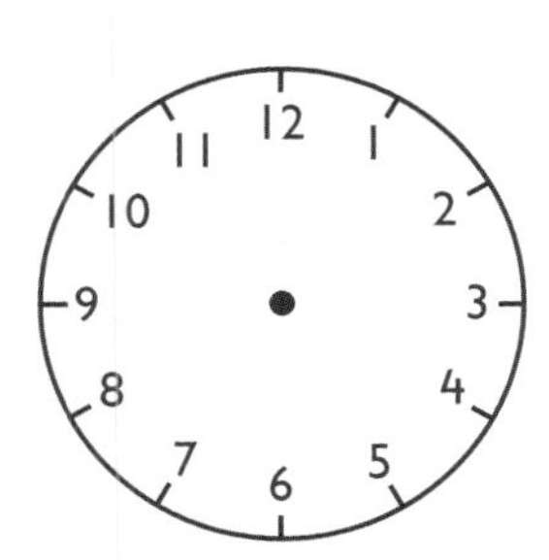

4. Which is the longest line?

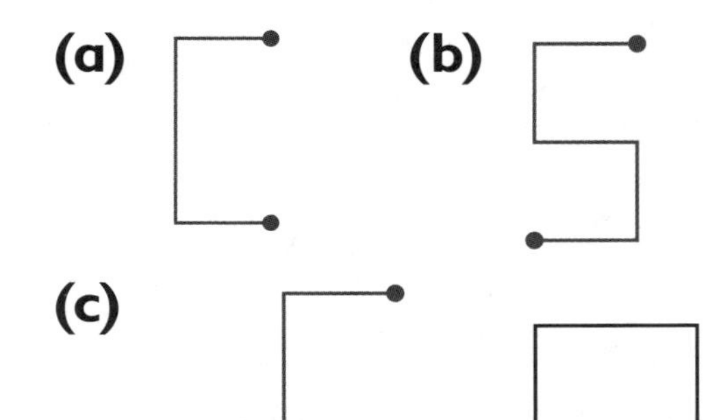

(c)

5. Shade three boxes that add up to 2.

1	2	0	1	3

6. Tina had 20 chocolates.
   She ate $\frac{1}{2}$ of them.
   How many did she have left?

7. 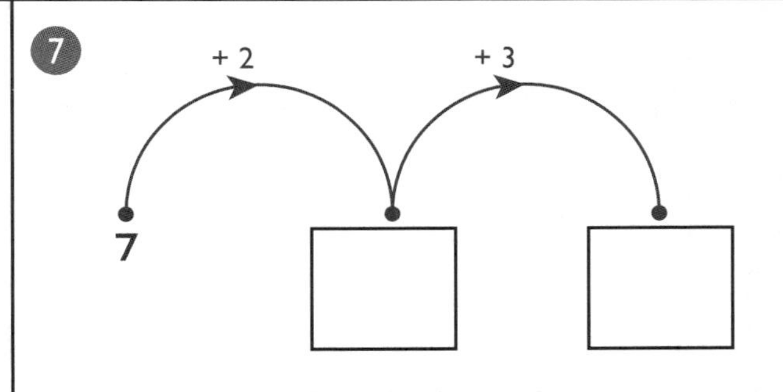

8. $6 + 6 + \square = 20$

9. $2 + \square = 5$

10. $9 - \square = 8$

Score

Test 16

1 12 units = ☐ ten + ☐ units

2 Two cats have ☐ legs.

3 Shade $\frac{1}{2}$ of this letter.

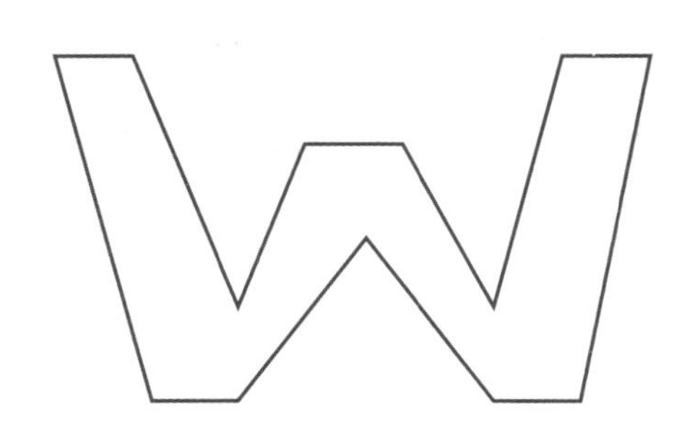

4 Which is the longest?

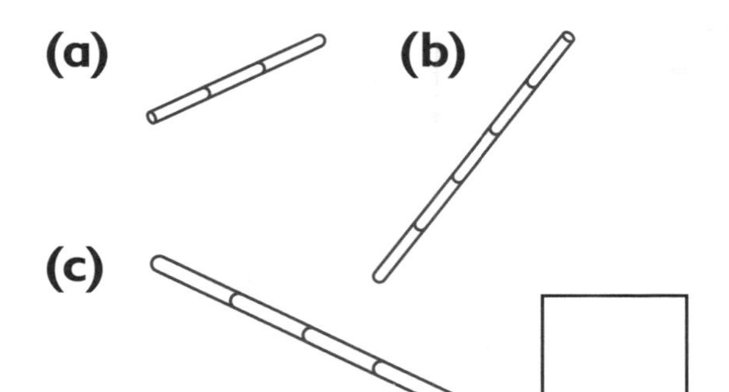

☐

5 Fill in the missing numbers.

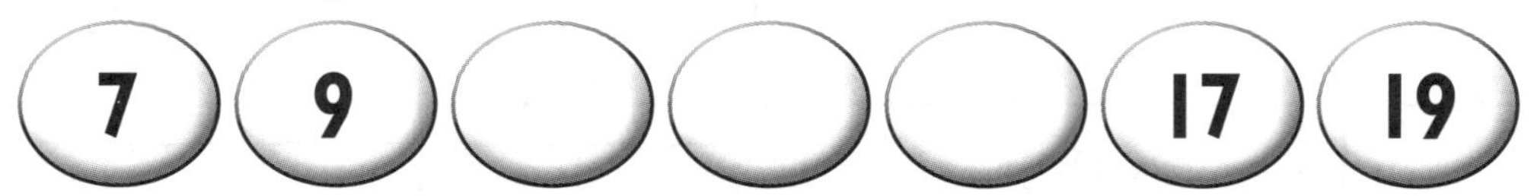

6 Fill in the missing numbers.

24, 26, ☐, ☐, ☐, 34

7 Ruby ate half of her chocolates and had 7 left.
How many did she have at the start?

☐ chocolates

8 Show 12:05

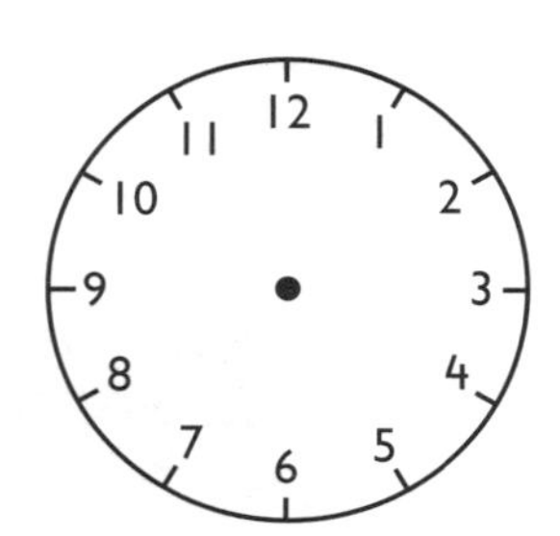

9 6 – ☐ = 6

10 1 + ☐ = 4

Score

Test 17

1. 7 − 3 = ☐

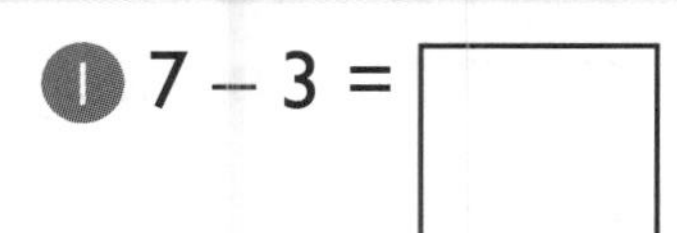

2. 24 = 2 tens and ☐ units

3. 10 minutes past 6 o'clock

☐ minutes past ☐ o'clock

4. A cube has ☐ faces.

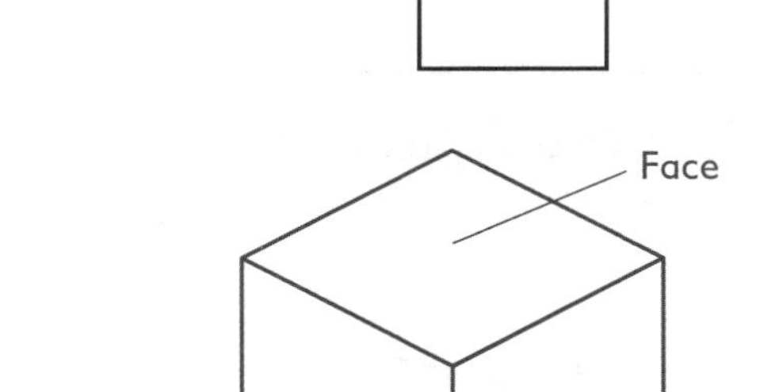

5. Shade three boxes that add up to 6.

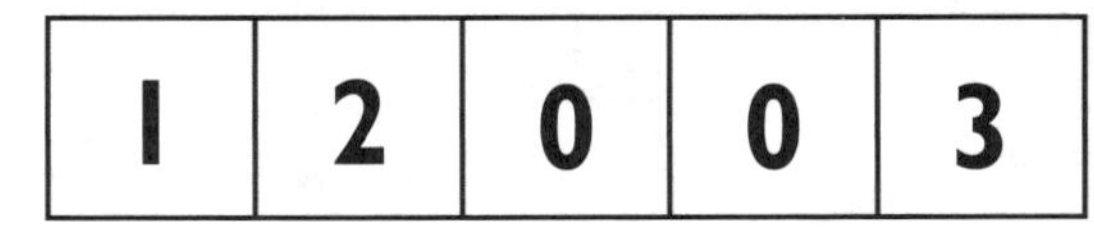

6. 10 pins are standing. If 4 are knocked down, how many will be left standing?

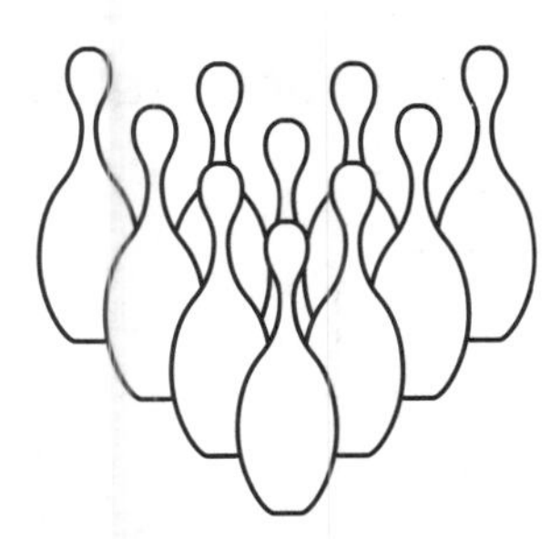

7. Shade $\frac{1}{2}$ of this letter.

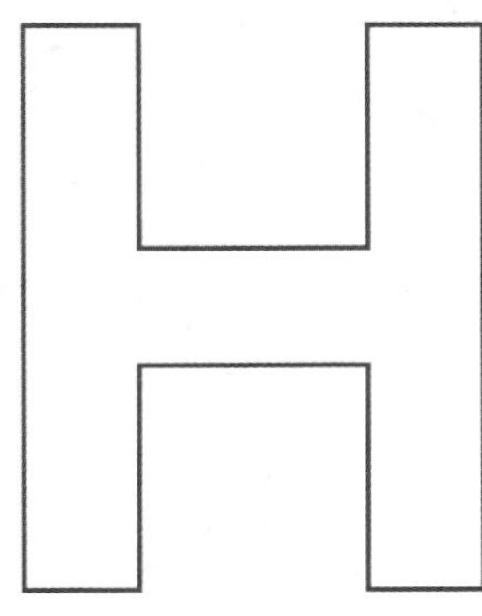

8. Draw the next two shapes.

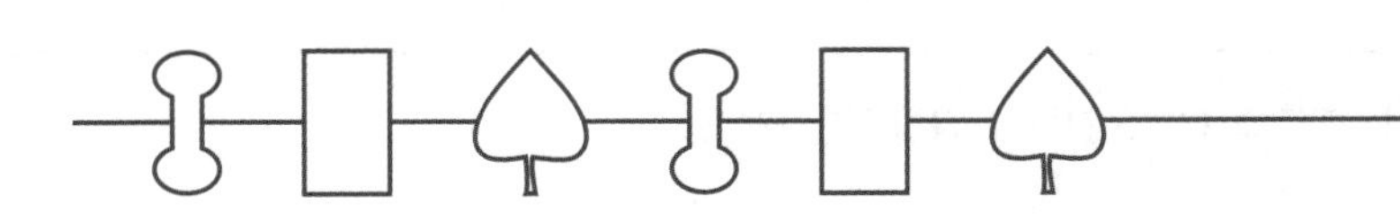

9. 12 − ☐ = 11

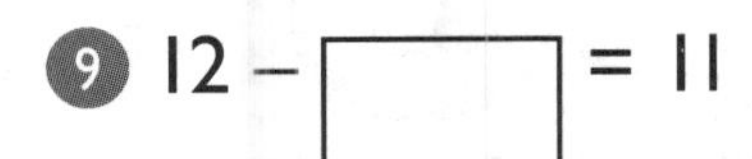

10. ☐ − 3 = 5

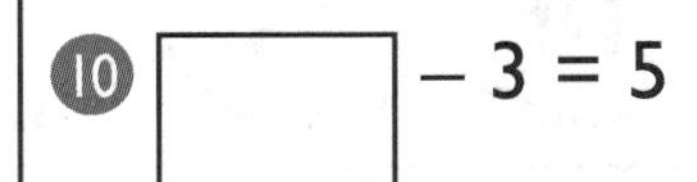

Score

Test 18

1 3 tens 5 units =

2

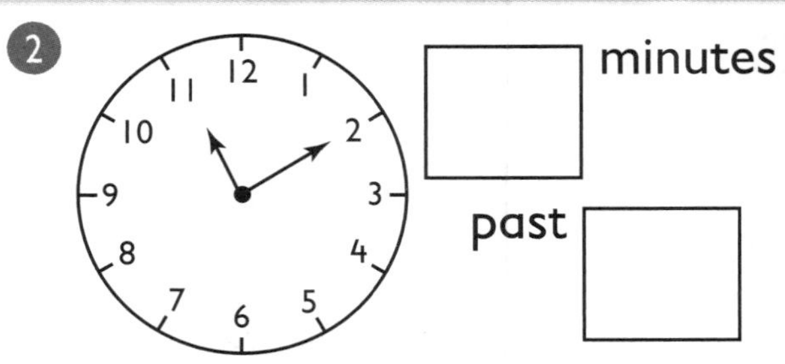

3 Which is the lighter, A or B?

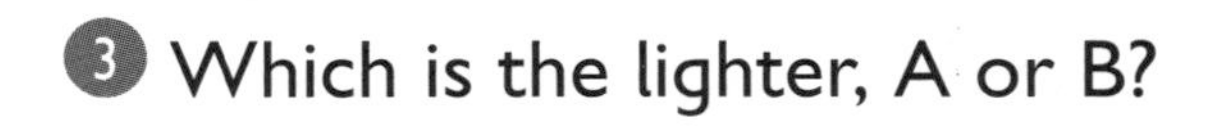

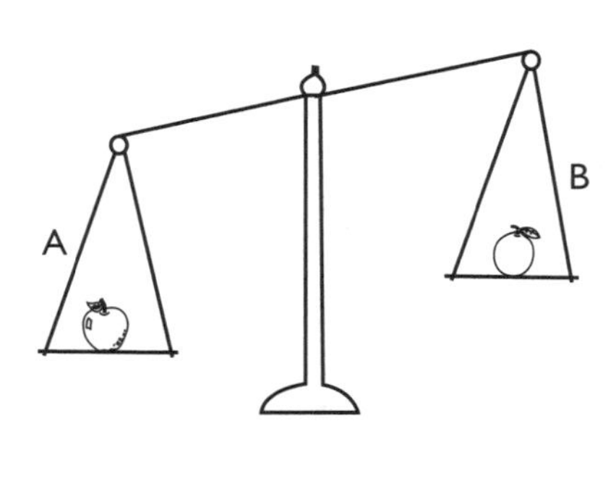

4 Which is the shortest?

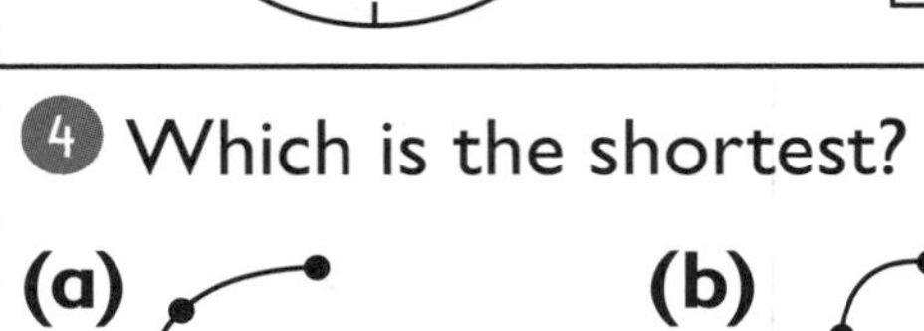

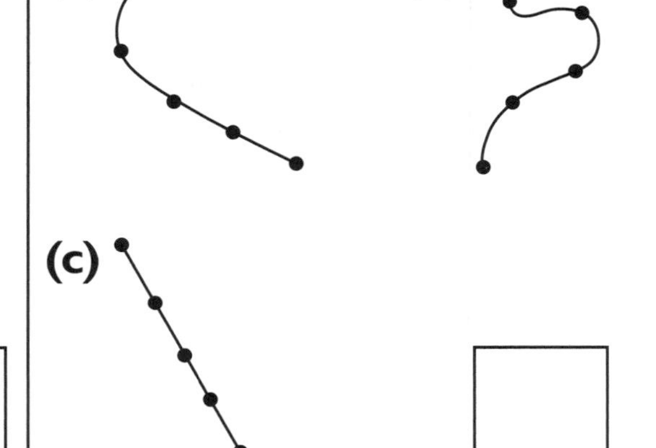

5 Draw the next two shapes.

6

+ 5 + 3

6

7 There are ☐ rectangles.

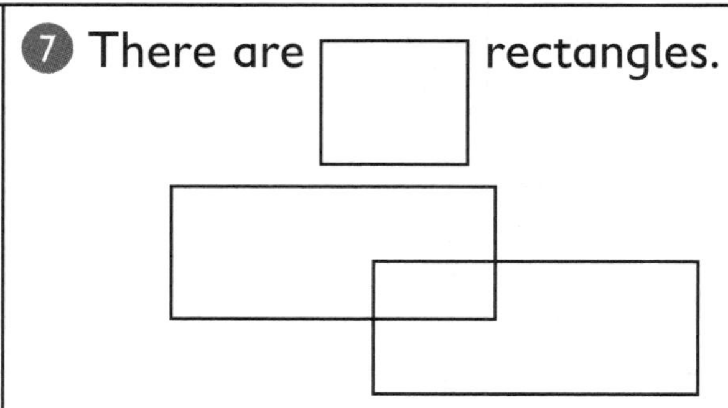

8 Fill in the missing numbers.

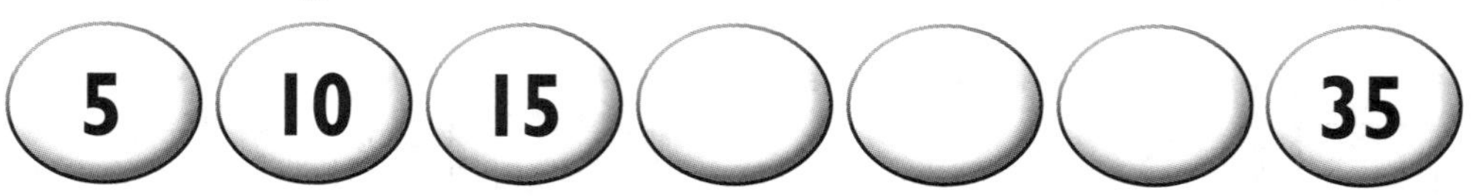

9 10 more than 80 is

10 8 – 5 =

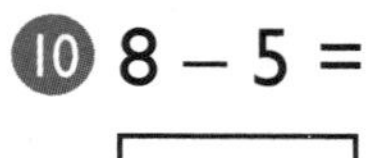

Test 19

1

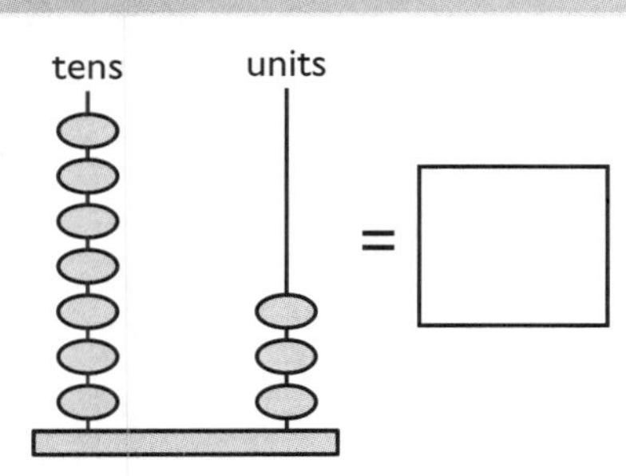

2

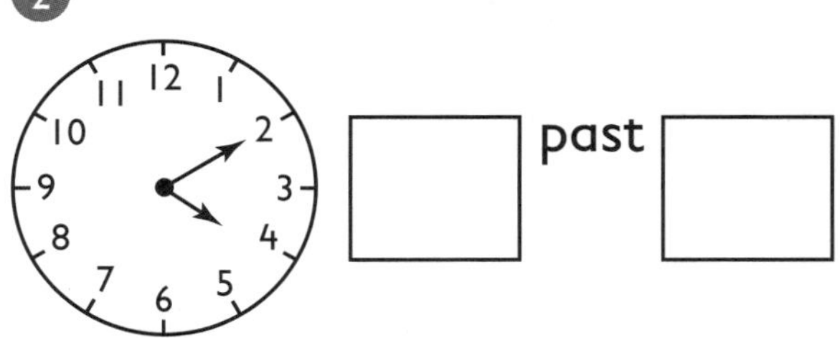

3 Shade $\frac{1}{2}$ of the letter.

4 Which is a cube?

(a) 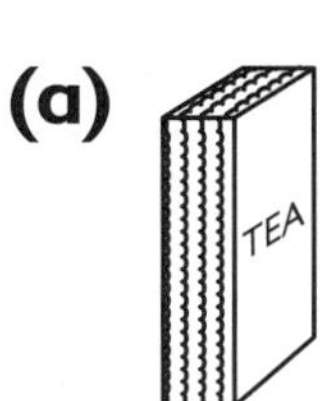

(b)  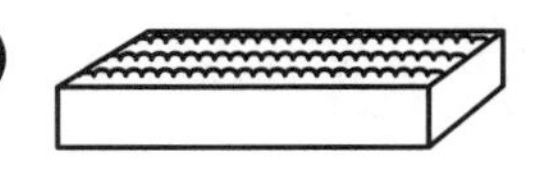

(c)

5 Which is the greatest number?  **(a)** 79 **(b)** 92 **(c)** 87

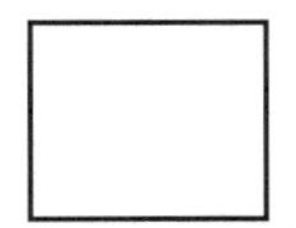

6 10 minutes past 6 o'clock is

10 minutes past 8 o'clock is

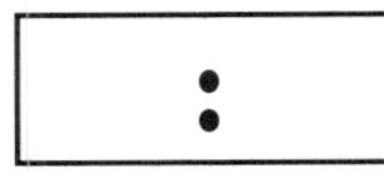

7 10 pins are standing.
If 2 are knocked down, then 6 are knocked down, how many will be left standing?

8 Count on in 10s.
Write the numbers.

50	60				100

9 35 − 2 = ☐

10 ☐ − 2 = 24

Score

Test 20

1 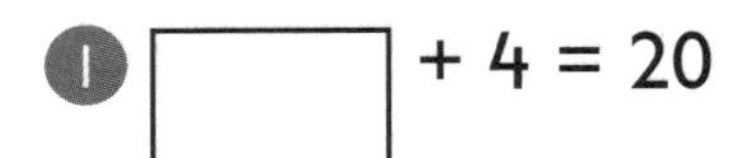 + 4 = 20

2 Show 86.

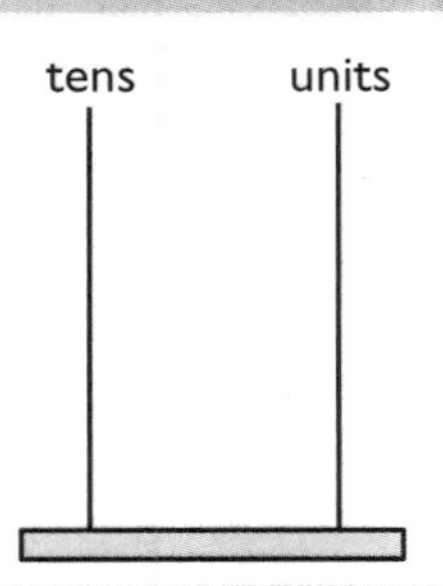

3 Which is a rectangle? 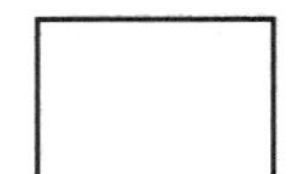

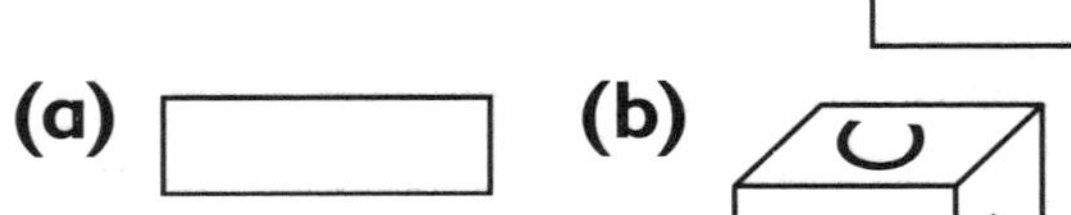

(c)

4 A cuboid has ☐ faces.

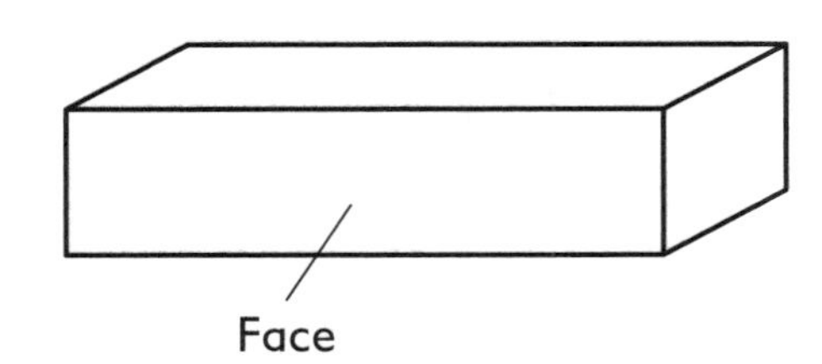

5 Draw the next 2 shapes.

6 Fill in the missing numbers.

97, 98, 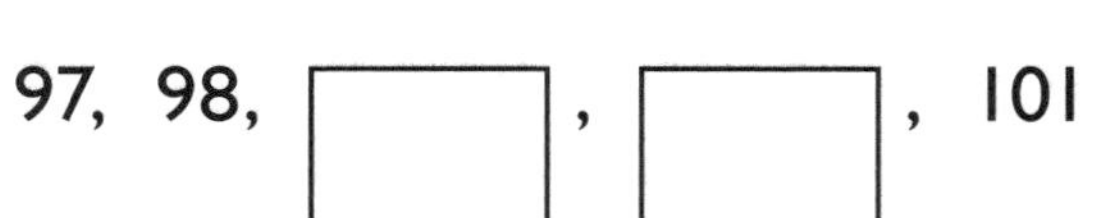 , ☐ , 101

7 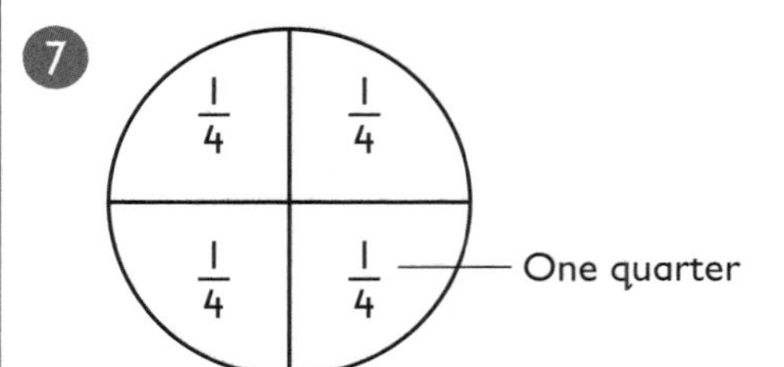

Shade one quarter of 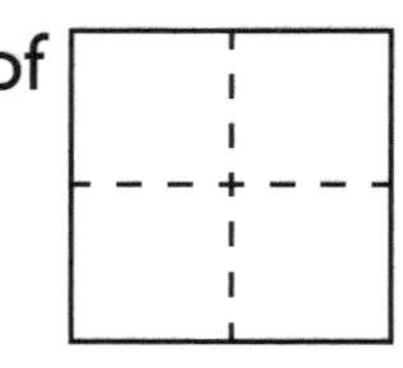

8 Fill in the missing numbers.

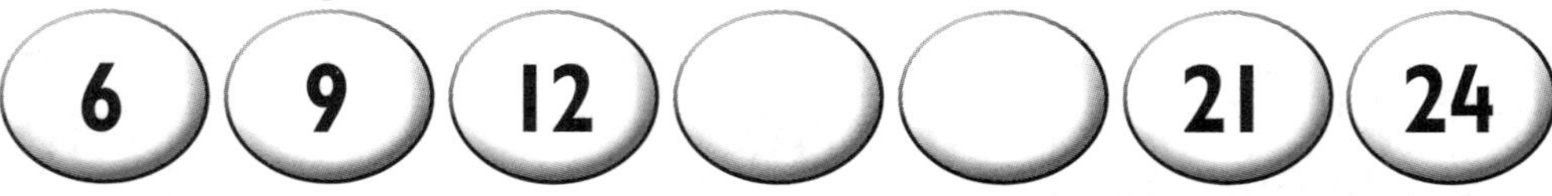

9 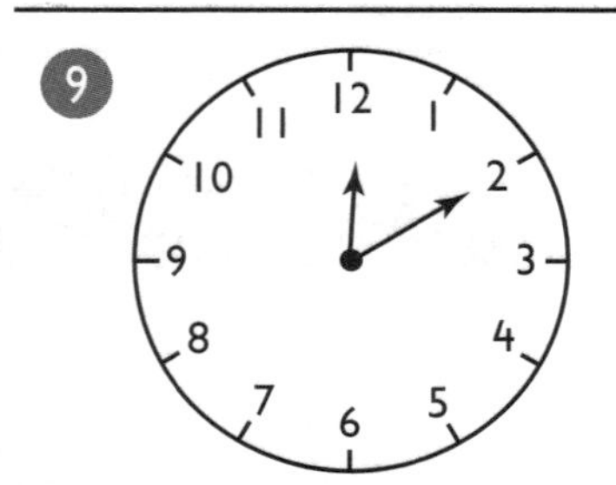

is ☐ :

10 14 − 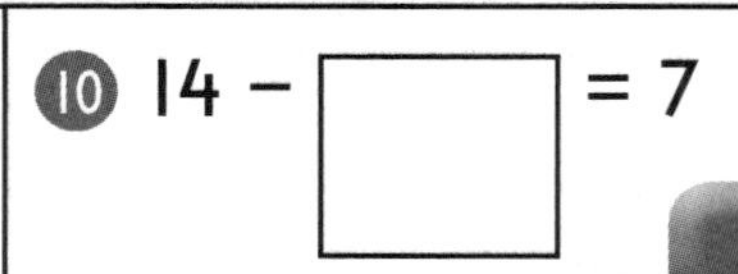 = 7

Score

Test 21

1 Which is the greatest?

65, 28, 96, 69

2 Fill in the missing numbers.

28, 29, 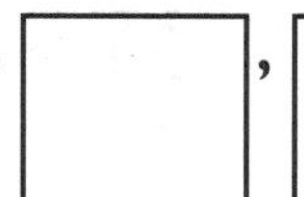, 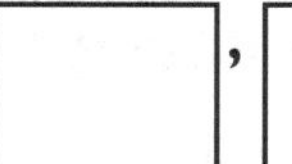, 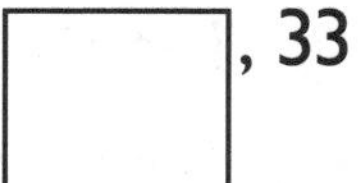, 33

3 Which rectangle has one quarter shaded?

(a) 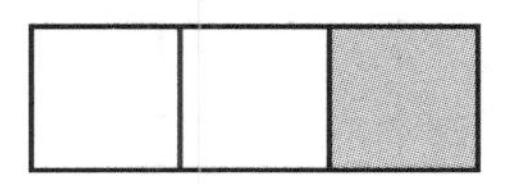

(b) 

(c) 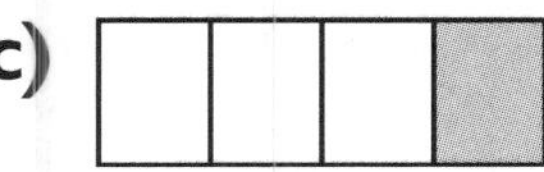

4

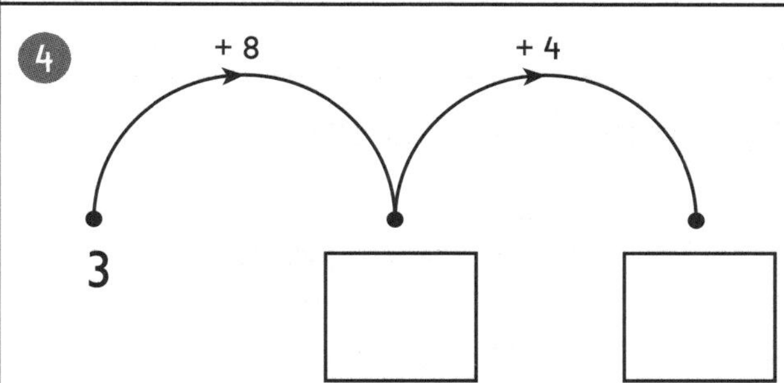

5 Fill in the missing numbers.

14  12 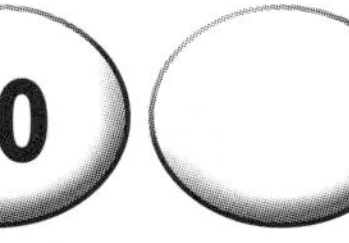 10 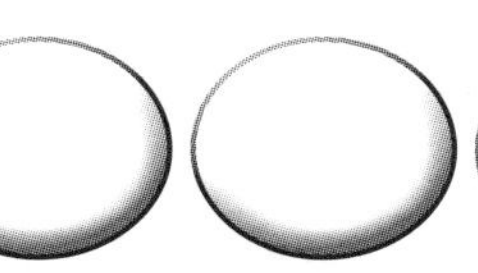 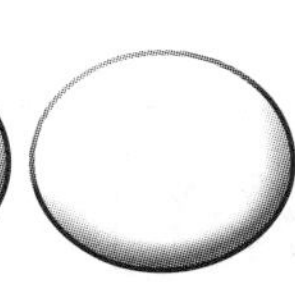  2

6 Fill in the number.

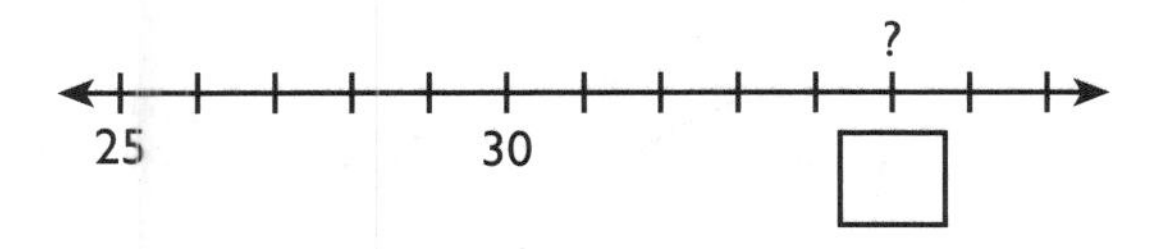

7 10 pins are standing.
If 7 are knocked down, then
2 more are knocked down,
how many will be left standing?

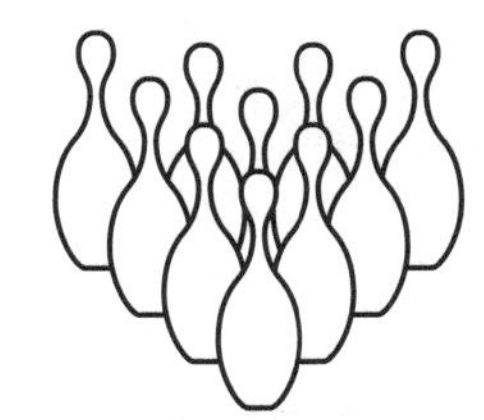

8 How many wheels are on three cars?
(Each car has four wheels.)

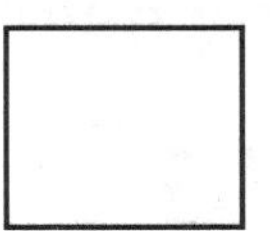

9 Show **6:10**

10 0 + 0 + 0 =

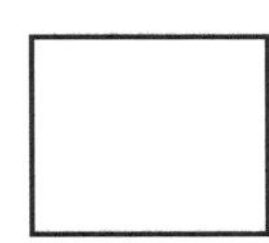

Score

Test 22

1 $>$ means greater than.

Is this correct? $24 > 19$
(yes or no)

2 Write the number one hundred and ten.

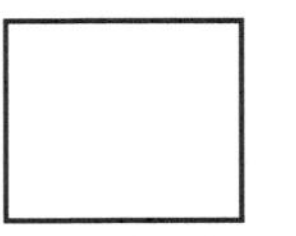

3 Shade $\frac{1}{4}$ of this set.

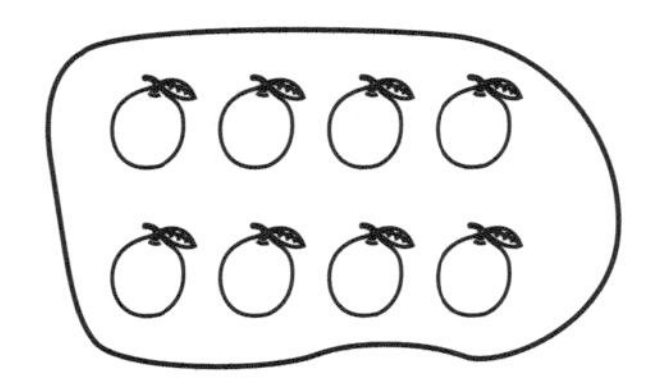

4 A cuboid has ☐ corners.

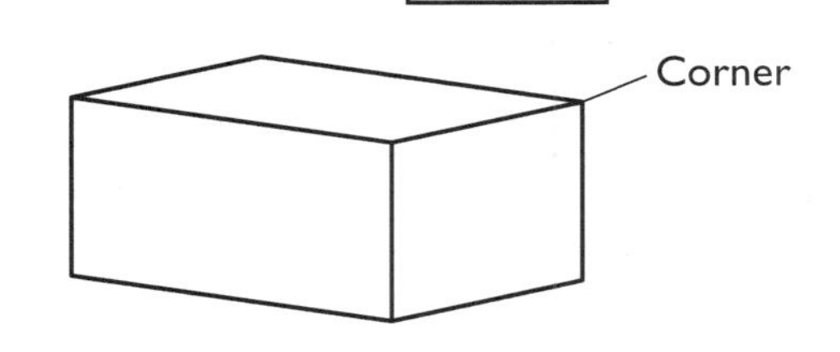

5 How many corners has a cube?

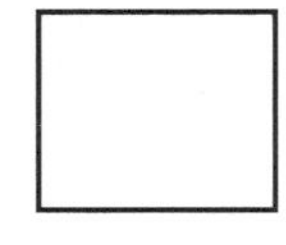

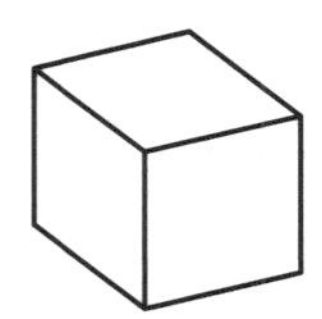

6 Show **10:10**

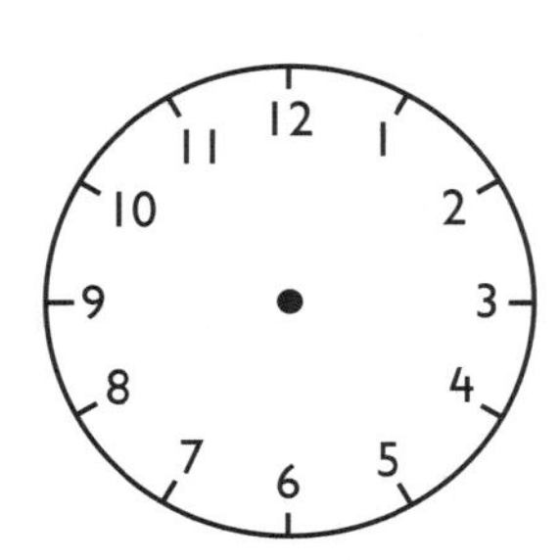

7 15 minutes past 8.

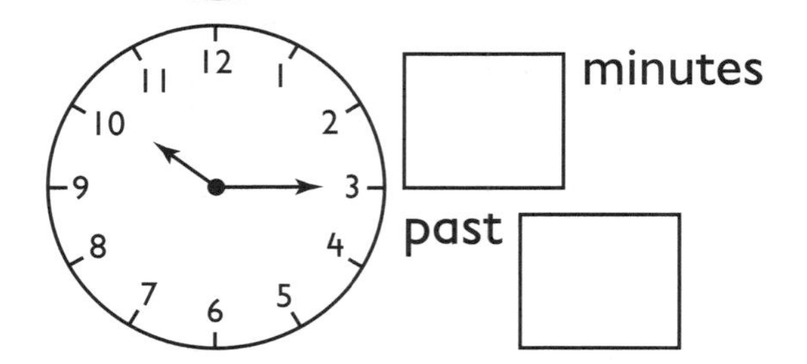

8 Fill in the missing numbers.

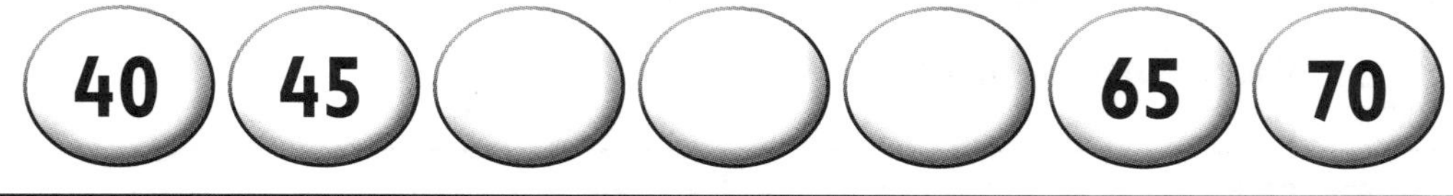

9 7 + 3 + 5 =

10 5 + 3 + 7 =

Score

Test 23

1. $\frac{1}{4}$ of 8 chocolates = ☐ chocolates.

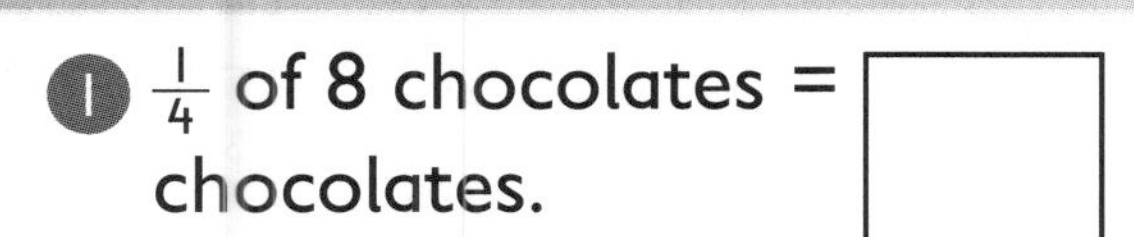

2. ☐ + 1 = 101

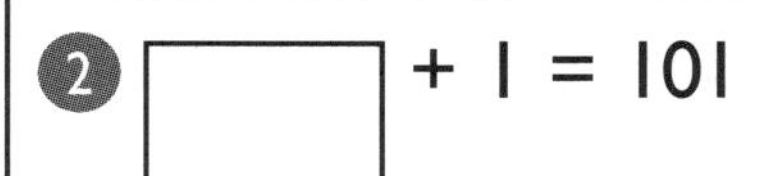

3. Which is a sphere?

(a)

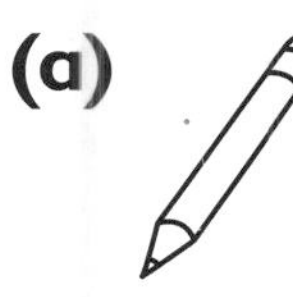

(b)

(c)

☐

4. Shade $\frac{1}{4}$ of this rectangle.

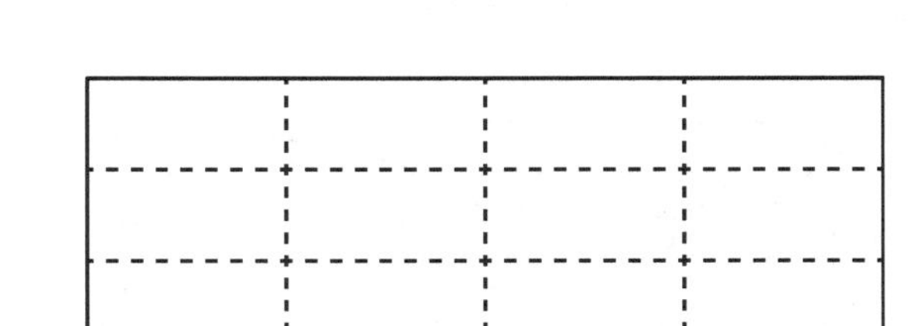

5. 10 − 1 = ☐

6. 8 + 8 = ☐

7.

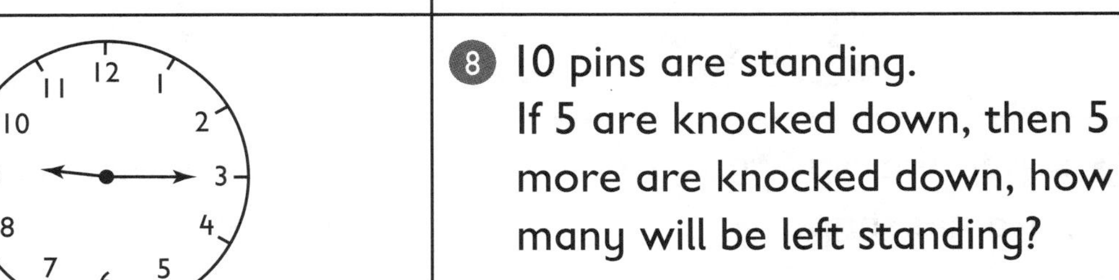

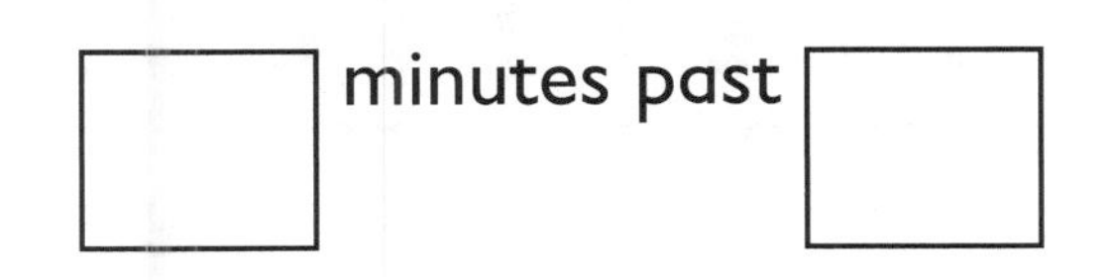

☐ minutes past ☐

8. 10 pins are standing.
If 5 are knocked down, then 5 more are knocked down, how many will be left standing?

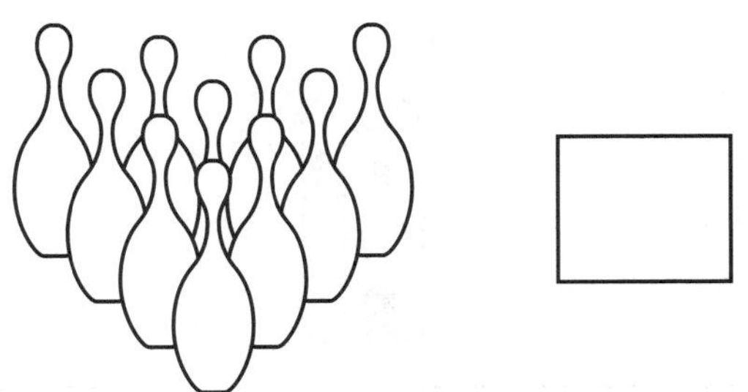

☐

9. The Moon is a ☐ (cube, sphere or cuboid).

10. 110 = 100 + ☐

Score

Test 24

1. 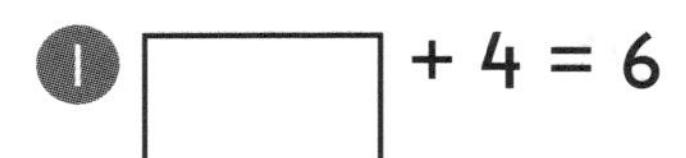 + 4 = 6

2. Which is a pyramid? 

(a) 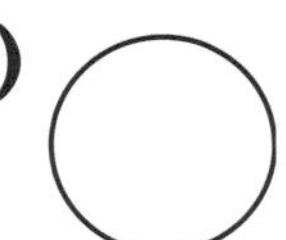 (b) 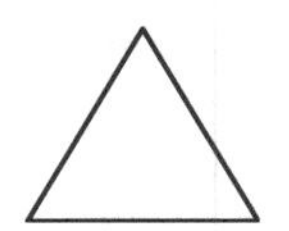 (c) 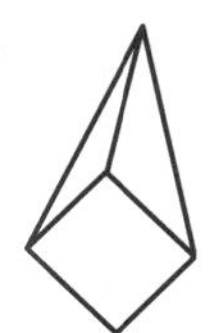

3. Shade $\frac{1}{4}$ of 12 toffees.

4. 

 minutes past

5. Fill in the missing numbers.

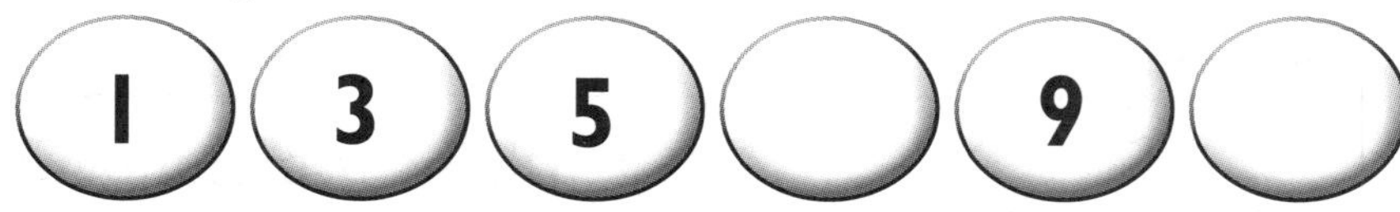

6. Shade the part that is a pyramid shape.

7. Shade two boxes that add up to 4.

8. Is 50 > 45? Write yes or no. 

9. 9 − ☐ = 6

10. 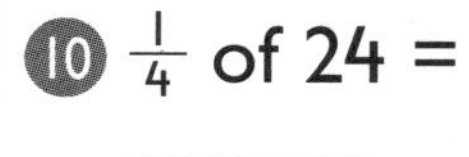 $\frac{1}{4}$ of 24 =

Score

Test 25

1. 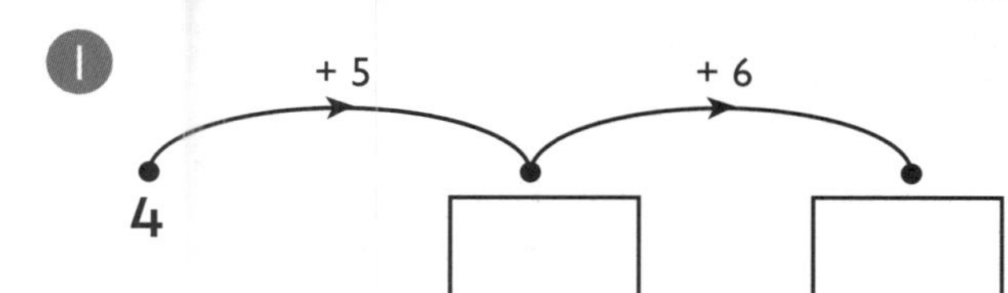

2. 10 less than 100 is ☐

3. Nita gave $\frac{1}{4}$ of her toffees to Charles.
How many did she have for herself?

4. Shade the part that is a cone.

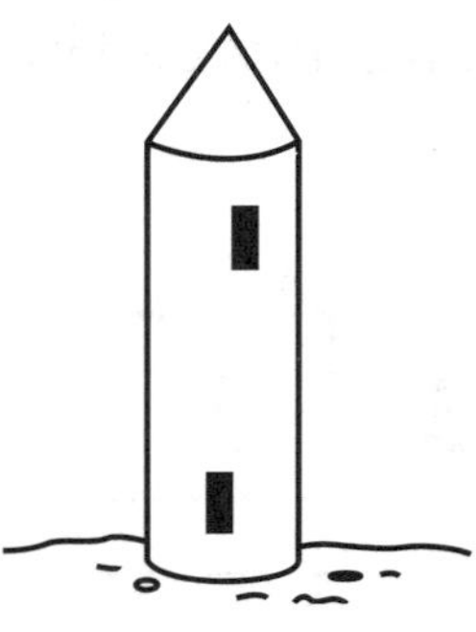

5. 15 minutes past 10 o'clock is

15 minutes past 6 o'clock is ☐ :

6. Draw the other half of this shape.

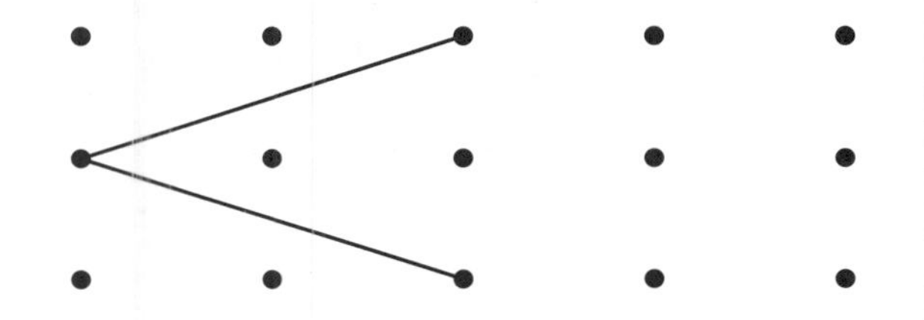

7. 10 pins are standing.
If 3 are knocked down, then 4 more are knocked down, how many will be left standing?

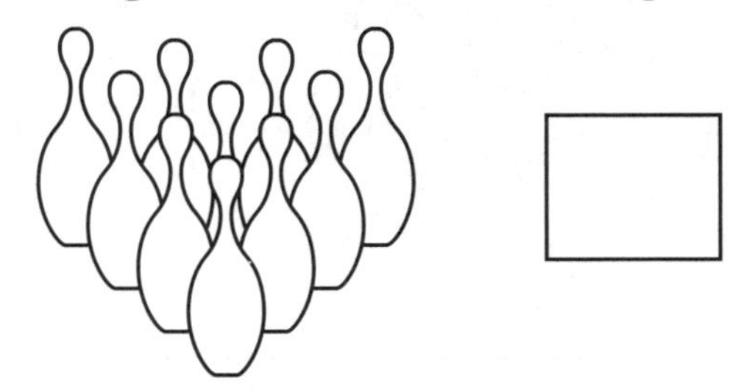

8. Fill in the missing numbers.

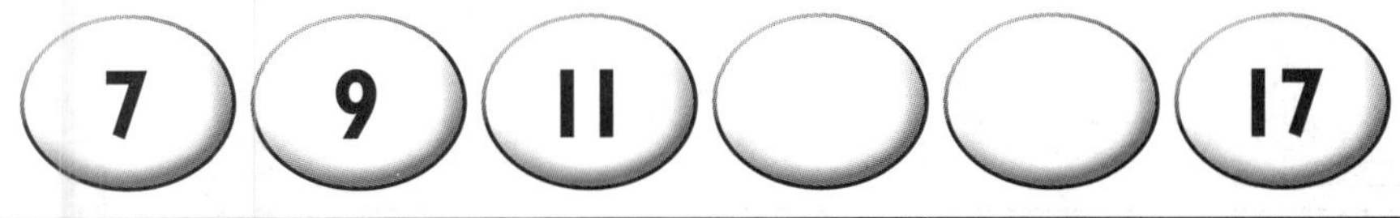

9. 17 − ☐ = 13

10. 12 + ☐ = 17

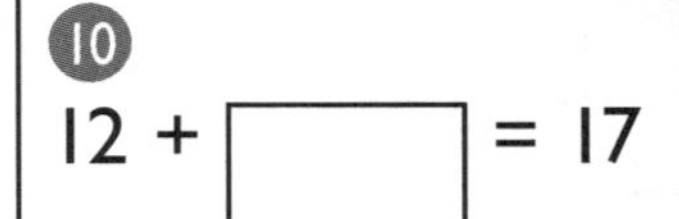

Score

Test 26

1 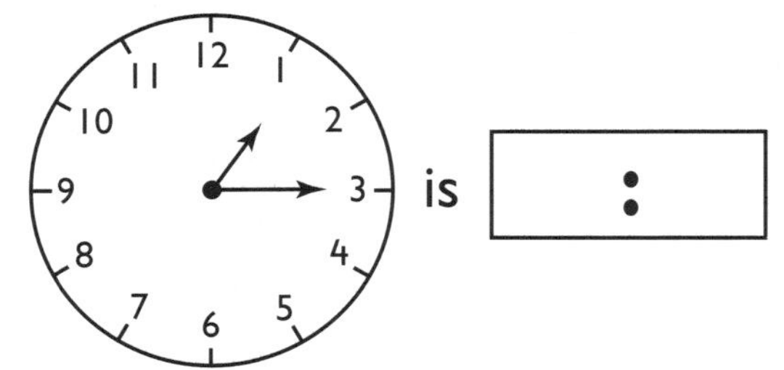

is [:]

2 Is 59 > 89?

Write yes or no.

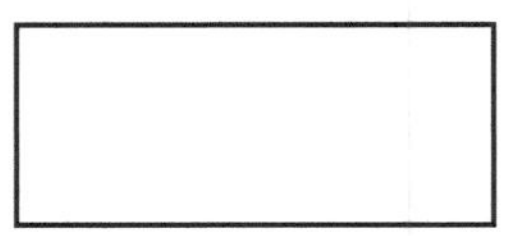

3 Shade two boxes that add up to 5.

1	3	3	4	5

4 Draw the other half of this shape.

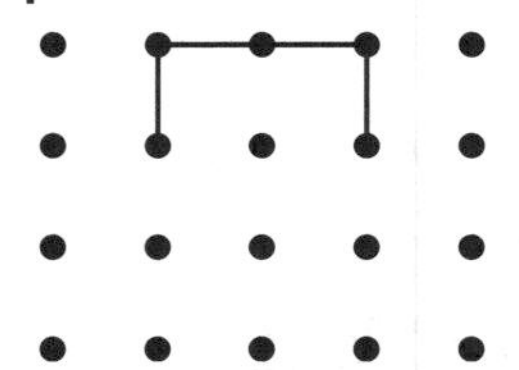

5 Fill in the missing numbers.

68 70 72 [ ] [ ] 78

6 10 pins are standing.
If 3 are knocked down, then
3 more are knocked down, how
many will be left standing?

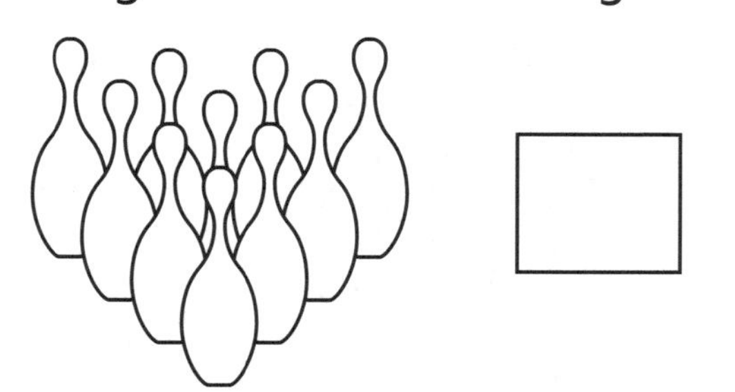

7 John is 9 years old.
His sister is 5 years older.
How old is his sister?

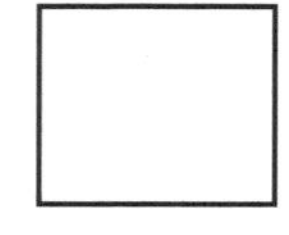

8 hundreds tens units

= [ ]

9 20 − [ ] = 15

10 39 − 2 = [ ]

Score

Test 27

1. $50 = 20 + 20 +$ ☐

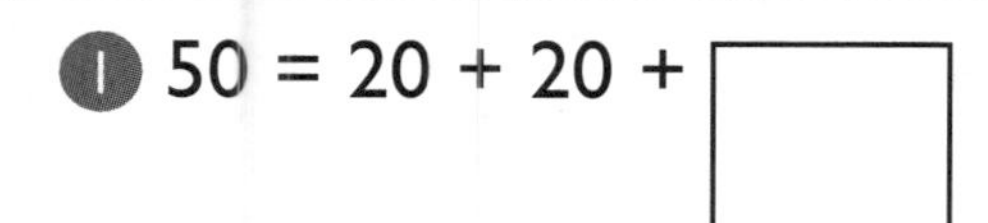

2. $\frac{1}{4}$ of $20 =$ ☐

3. Draw the other half of this shape.

4. Anhil is 7 years old.
   Diane is 13.
   How much older is Diane?
   ☐ years

5. Fill in the missing numbers.

21 24 ☐ ☐ 33 36 ☐ ☐

6. Draw a ring around the greatest number.

   119

   91

   179

7. $50 = 10 + 20 +$ ☐

8. Shade $\frac{1}{2}$ of this circle.

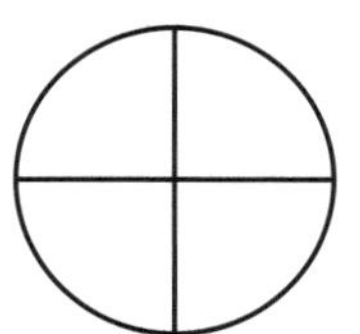

9. Show 5:15

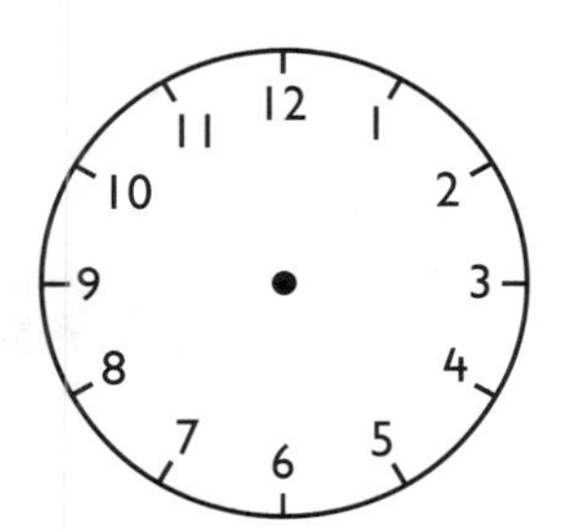

10. ☐ $+ 5 = 25$

Score

Test 28

1. 12 + ☐ = 20

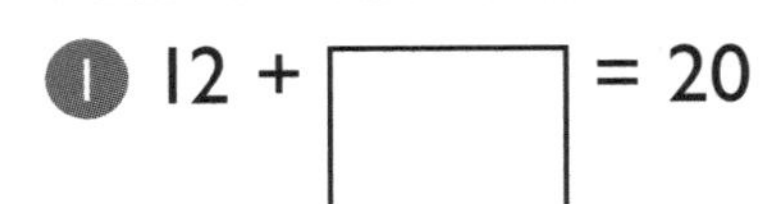

2. $\frac{1}{4}$ of 40 = ☐

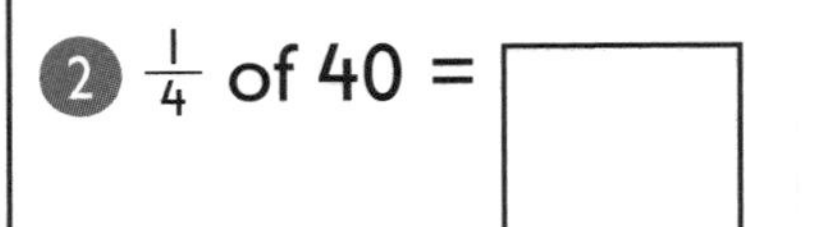

3. Which shape shows 2 quarters shaded?

(a) (b)

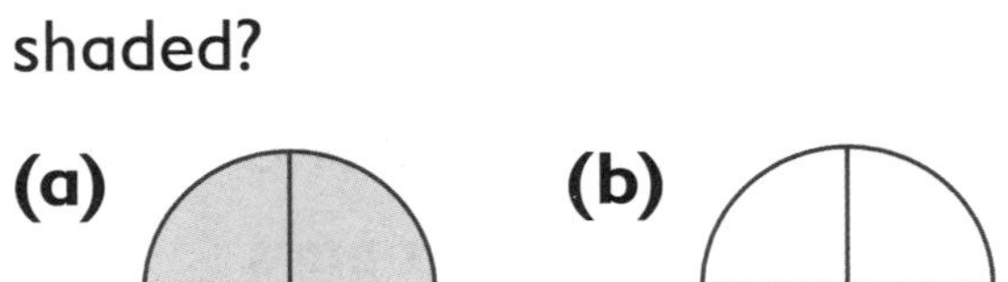

(c)

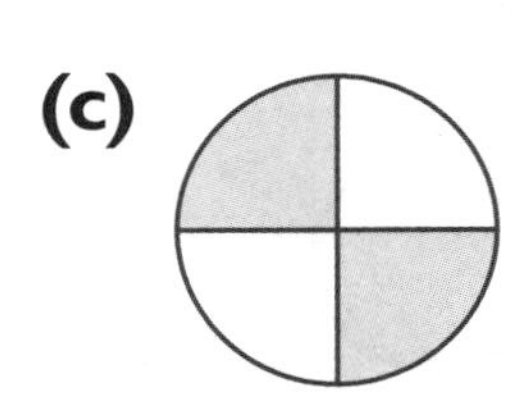

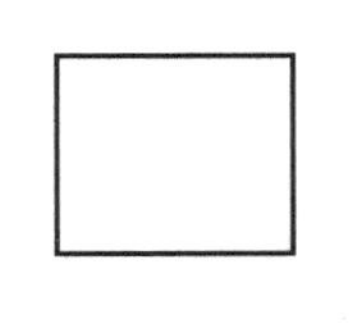

4. 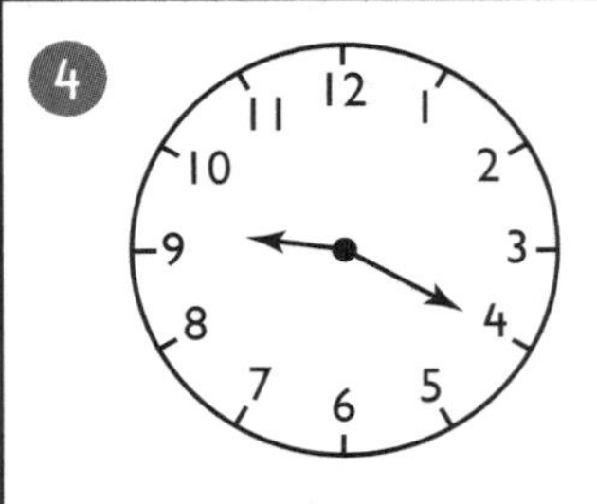

20 minutes past 9 o'clock

☐ minutes past

☐ o'clock

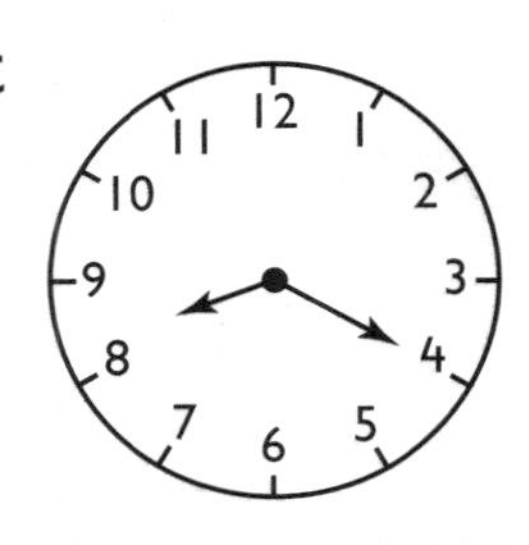

5. Fill in the missing numbers.

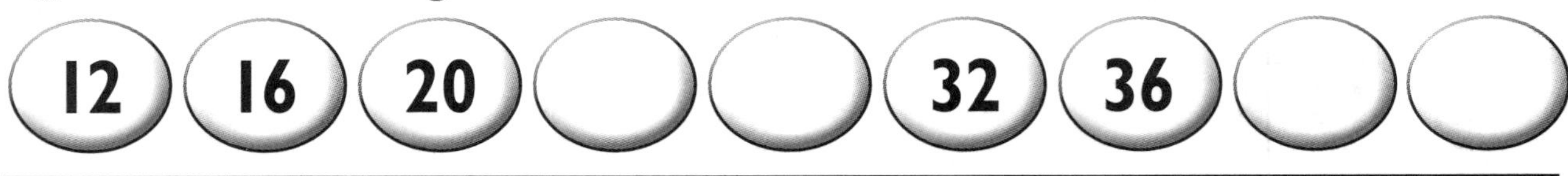

6. Tony had 50 peanuts.
He ate 20.
How many did he have left?

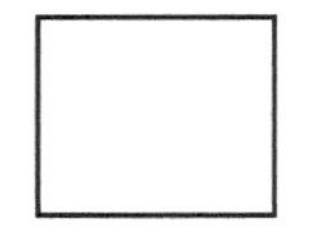

7. 

20 past 7 is **7:20**

20 past 3 is ☐ : ☐

8. ☐ + 4 + 4 = 20

9. 25 = 10 + ☐

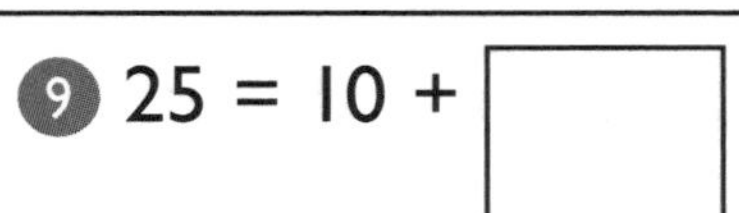

10. 20 − ☐ = 8

Score

Test 29

1. 19 − 6 =

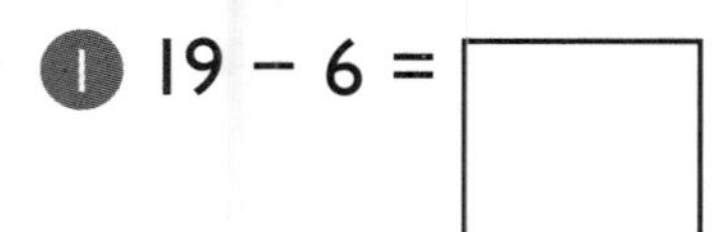

2. 27 = 20 +

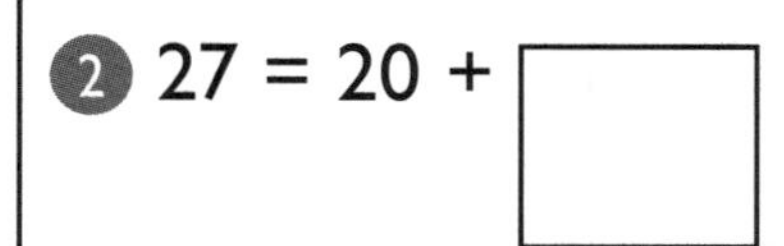

3. ☐ minutes past ☐

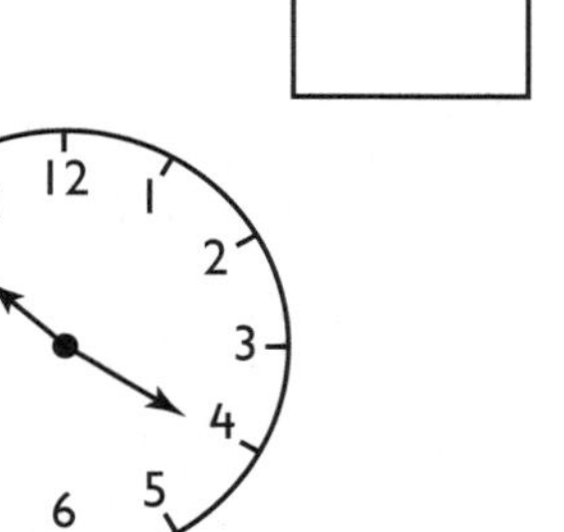

4. Shade $\frac{3}{4}$ of this shape.

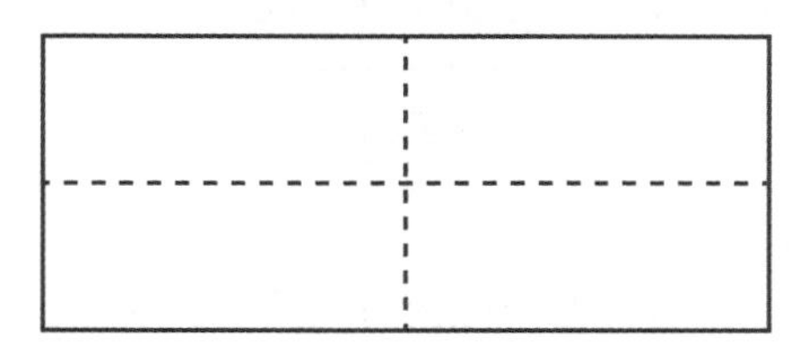

5. Fill in the missing numbers.

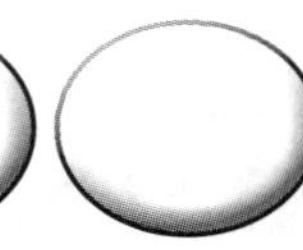

20 18 16 ☐ 12 ☐

6. Fill in the missing number.

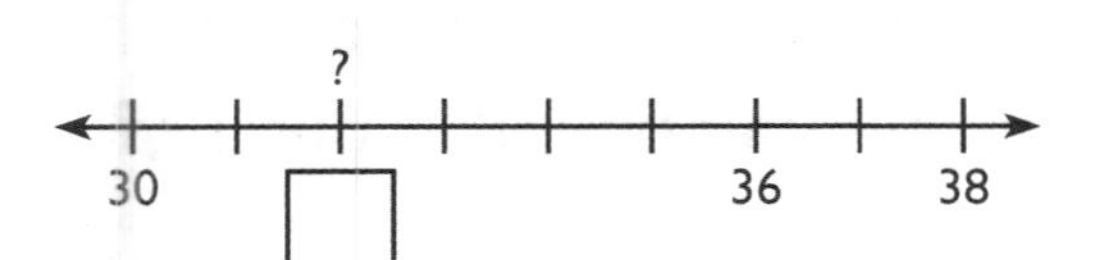

7. 10 pins are standing.
If 5 are knocked down, then 2 more are knocked down, how many will be left standing?

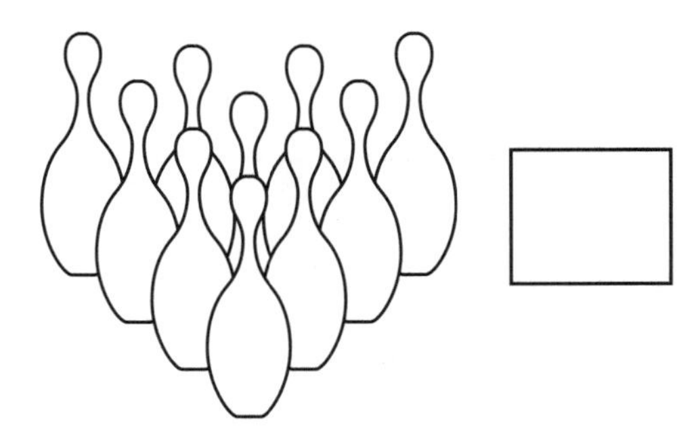

8. Half of 16 is

9. 18 − ☐ = 7

10. 20 − ☐ = 7

Score

Test 30

1. $29 - 7 =$ 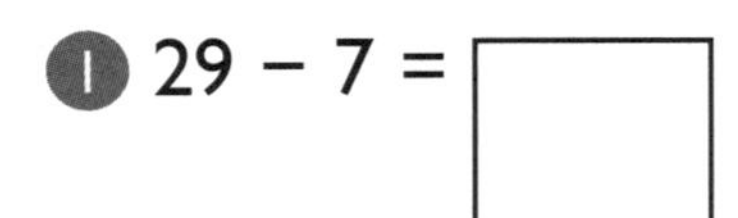

2. Is $29 > 20$?
   Write yes or no.
   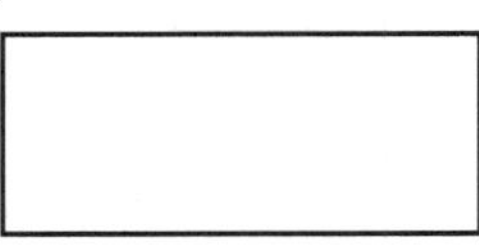

3. 20 minutes past 7 o'clock is

   7:20

   20 minutes past 1 o'clock is

   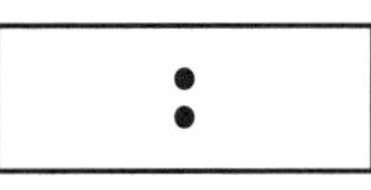

4. How many chocolates are there in $\frac{1}{2}$ of the box?

   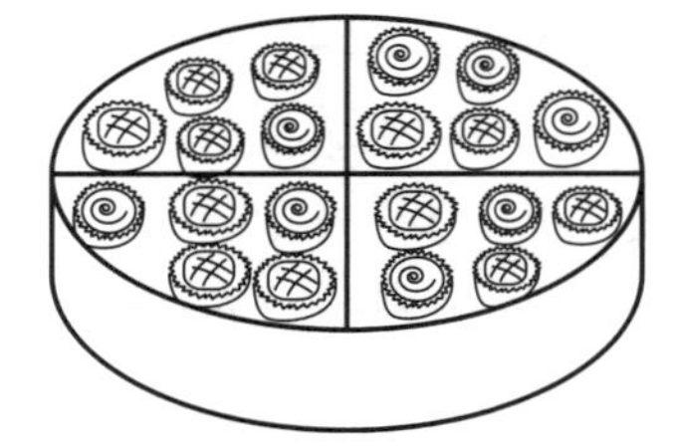
   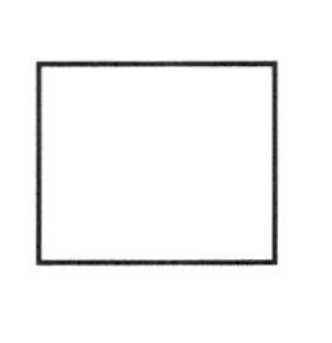

5. 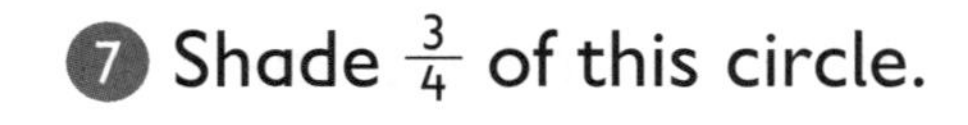
   $52 = \square + 2$

6. $20 - 3 = \square$

7. Shade $\frac{3}{4}$ of this circle.

   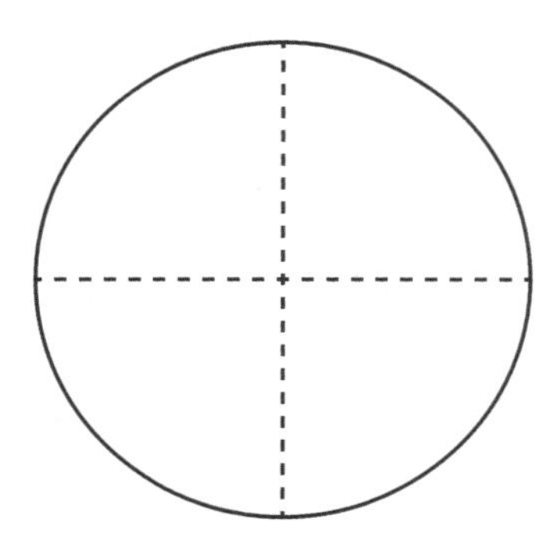

8. 10 pins are standing.
   If 6 are knocked down, then 4 more are knocked down, how many will be left standing?

   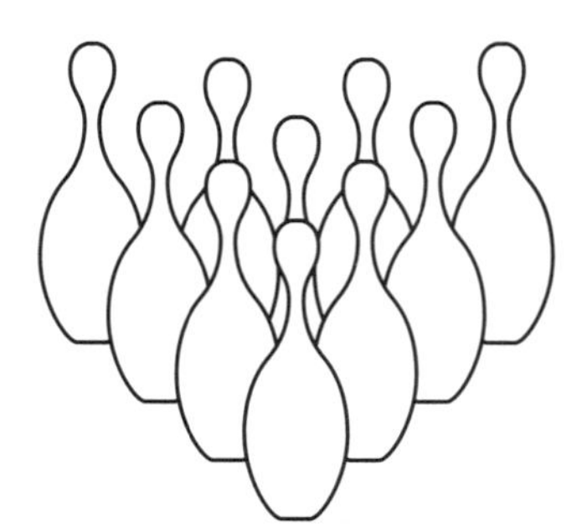

9. Fill in the missing numbers.

   5, 10, 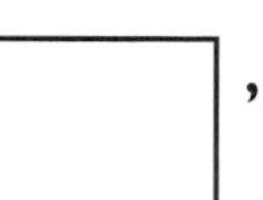 , $\square$, $\square$, 30

10. $48 = 40 + \square$

Score

Test 31

1. Show 726 on this abacus.

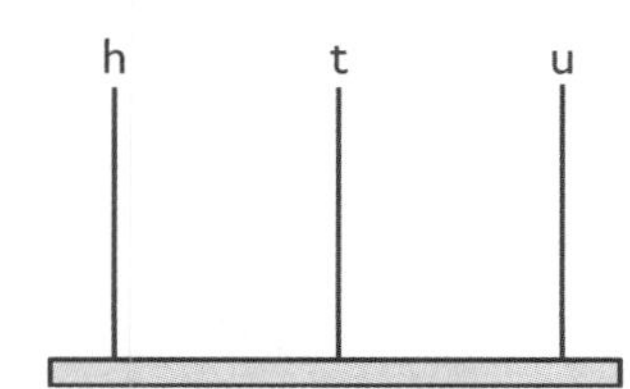

2. 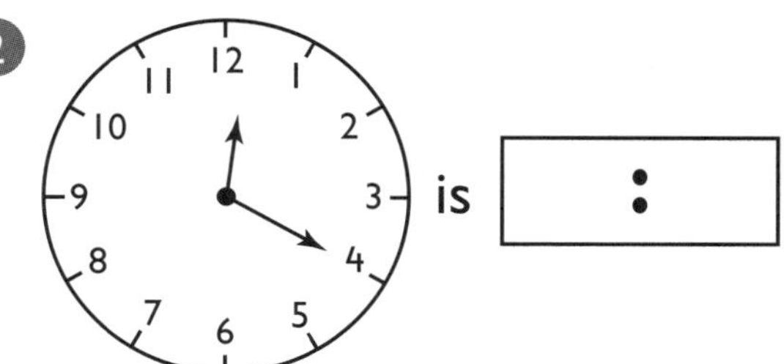

3. 10 pins are standing.
If 4 are knocked down, then 3 more are knocked down, how many will be left standing?

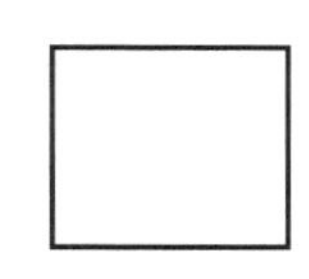

4. How many chocolates are there in $\frac{3}{4}$ of this box?

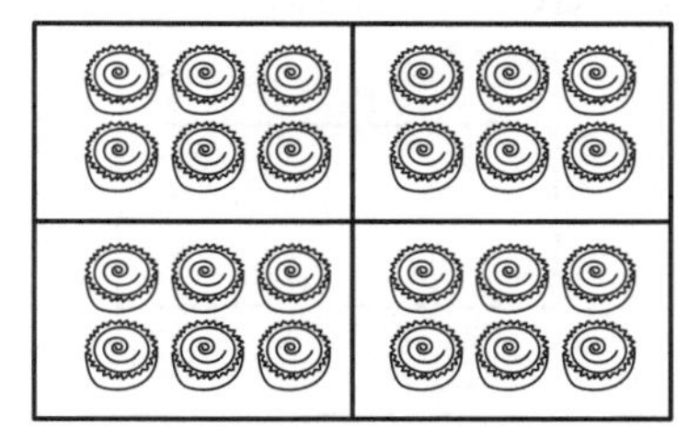

5. Fill in the missing numbers.

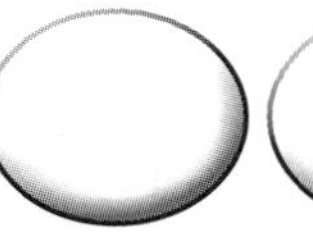

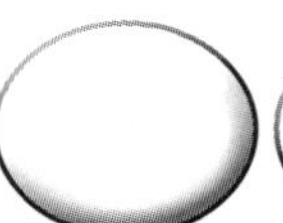

6 | 12 | 18 | 24 | 30 | 36 | | |

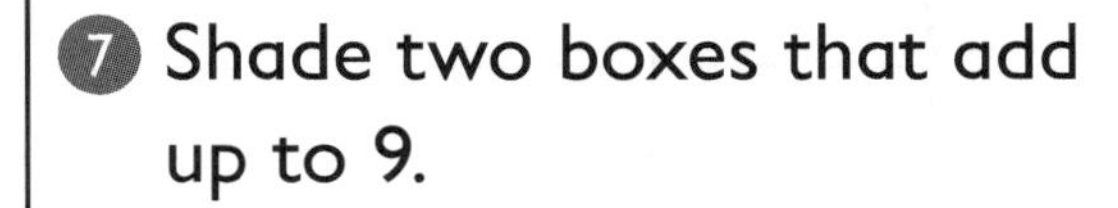

6. Shade $\frac{3}{4}$ of this circle.

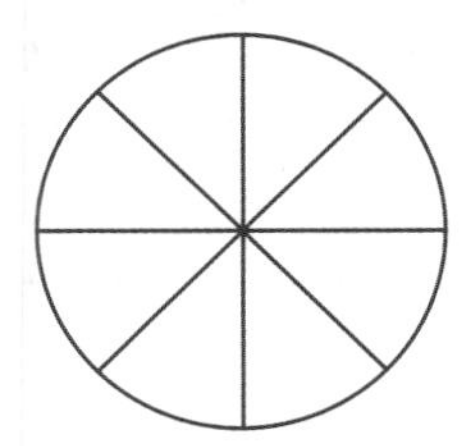

7. Shade two boxes that add up to 9.

8. Fill in the missing numbers.

9. 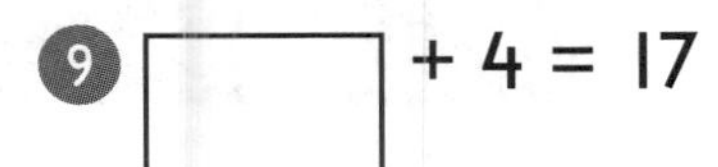 + 4 = 17

10. 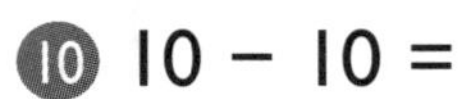 10 − 10 =

Score

Test 32

1. 9 − 8 =

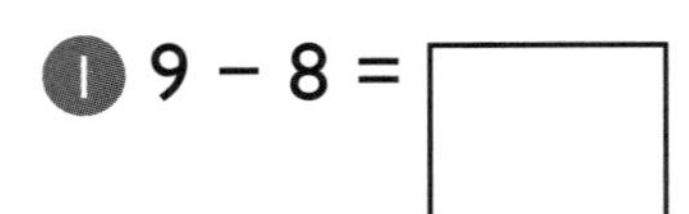

2. 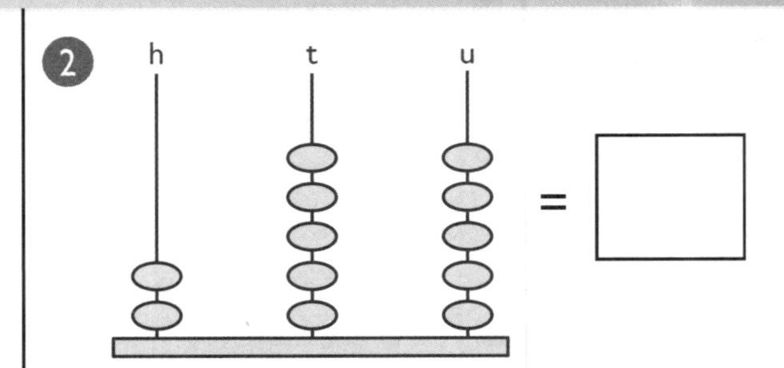

3. Is 36 > 63?

   Write yes or no.

4. Show 20 minutes past 10.

5. 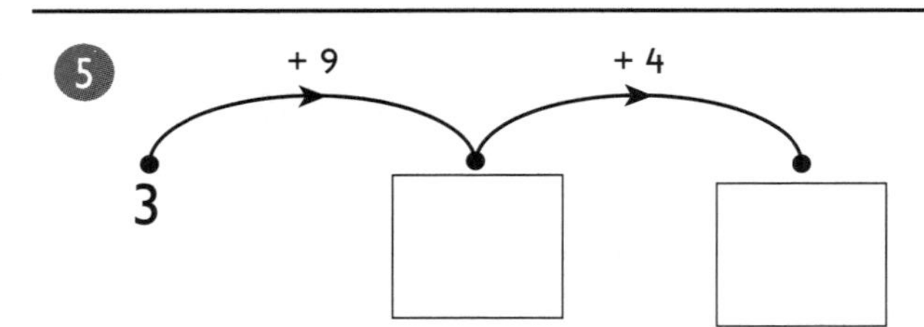

6. Shade $\frac{3}{4}$

   $\frac{1}{4}$ $\frac{1}{4}$ $\frac{1}{4}$ $\frac{1}{4}$

7. Shade two boxes that add up to 8.

1	2	3	4	5

8. Fill in the missing numbers.

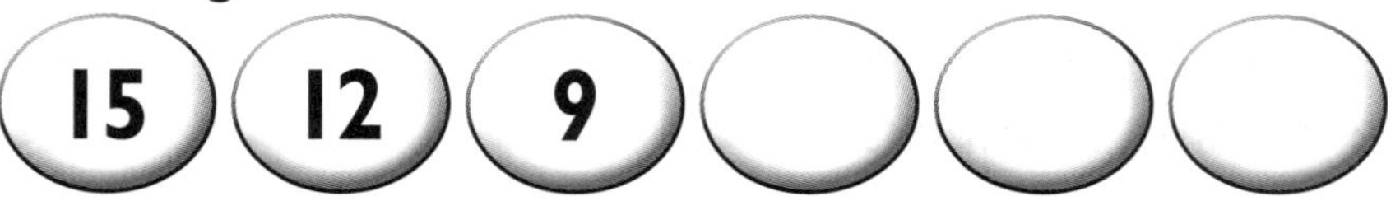

9. Show 253.

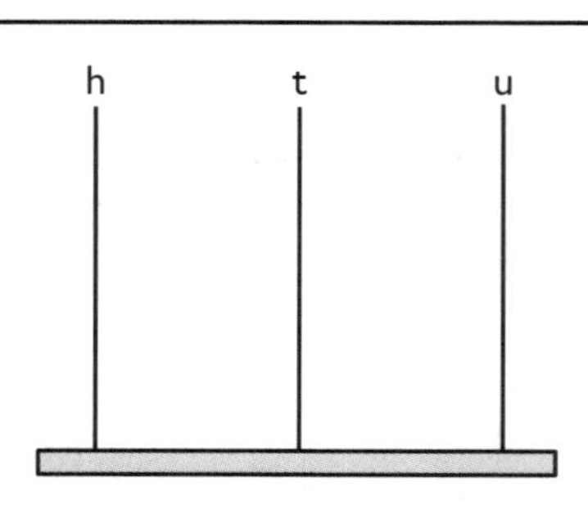

10. 20 − ☐ = 6

Score

Test 33

1. Count on in 4s.
   Write the numbers.

0	4				20

2. Four chairs have ☐ legs.

3. There are 21 children in a class.
   9 are boys.
   How many girls are there?

☐ girls

4. Shade two boxes that add up to 7.

1	2	4	4	5

5. Fill in the missing numbers.

36 34 32 30

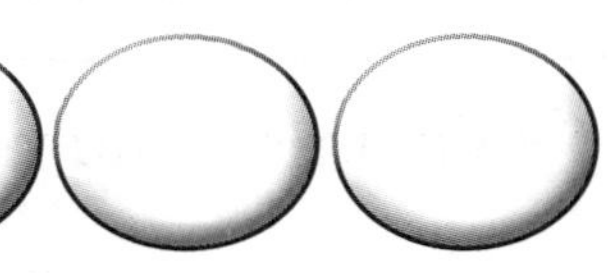

6. Shade $\frac{3}{4}$ of this rectangle.

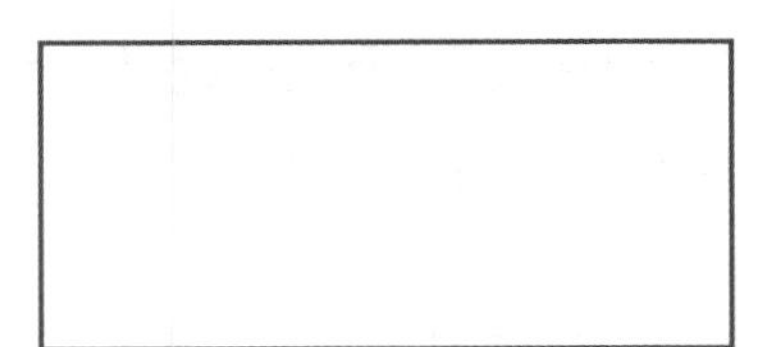

7.

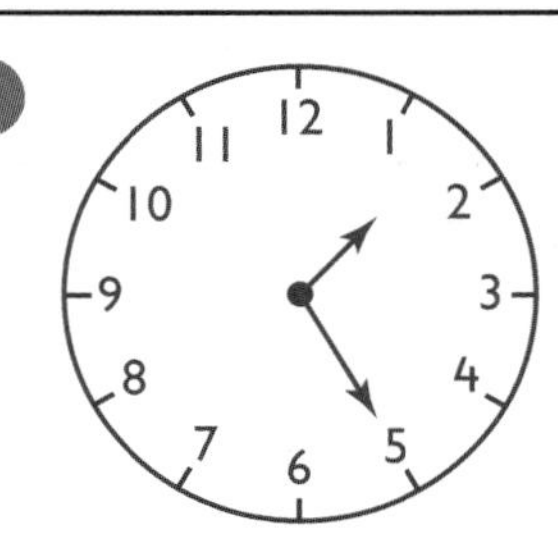

25 minutes past 1 o'clock

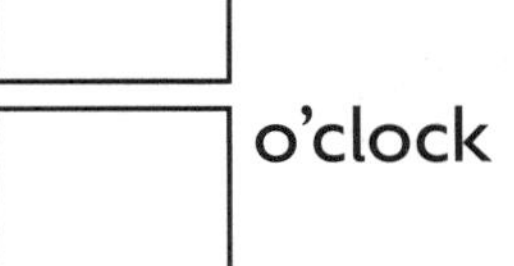

☐ minutes past

☐ o'clock

8. Draw the two missing shapes.

9. 3 hundreds 0 tens 3 units

= ☐

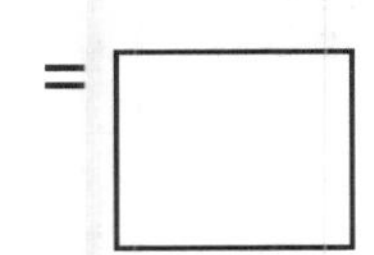

10. 9 + 2 =
    5 + ☐

Score

Test 34

1 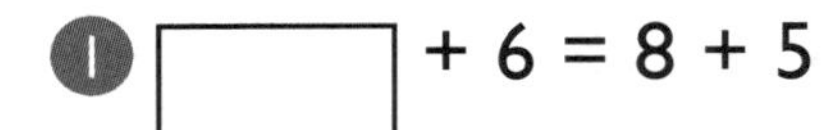 + 6 = 8 + 5

2 Shade $\frac{1}{4}$ of this shape.

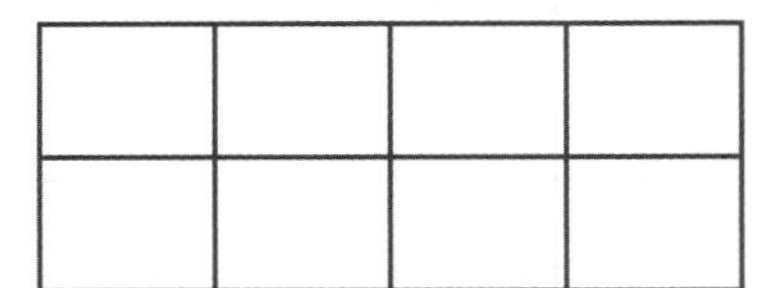

3 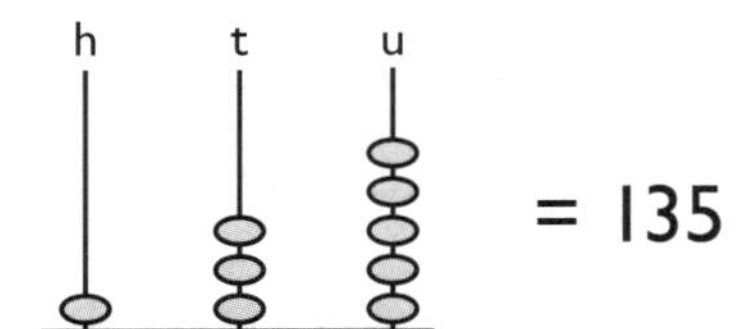

= 135

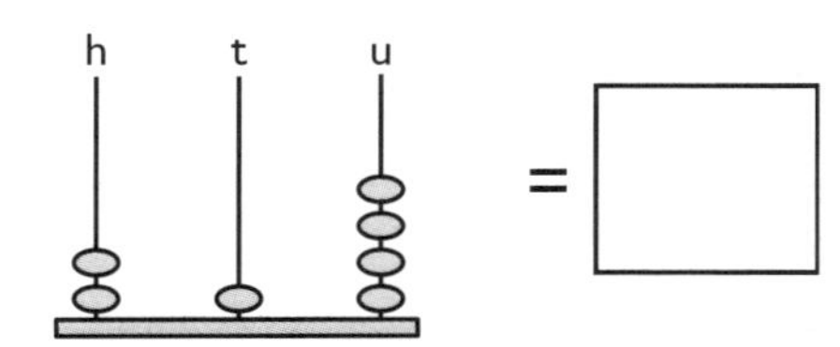

= ☐

4 Shade three boxes that add up to 6.

5 Fill in the numbers.

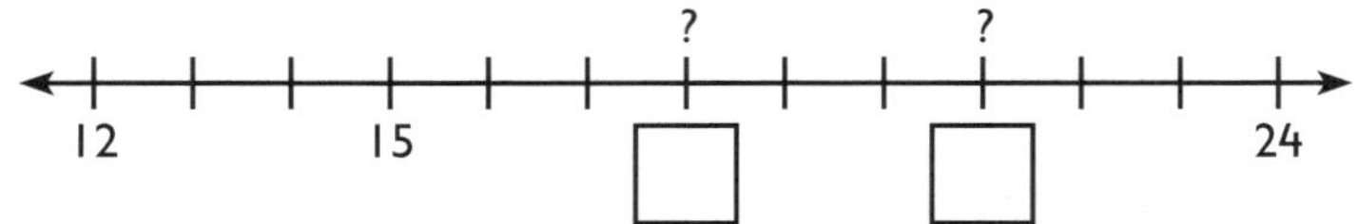

6 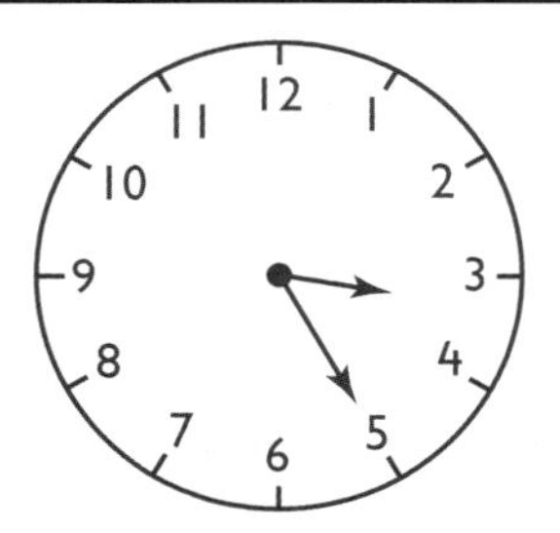

☐ minutes past ☐

7 Richard ate $\frac{1}{4}$ of a cake each day.
How long did it take him to eat half of the cake?

 days

8 Darren is 19 years old. Clare is 7.
How much younger is Clare?

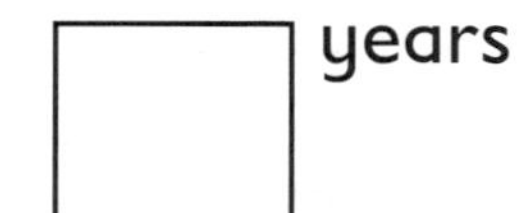 years

9 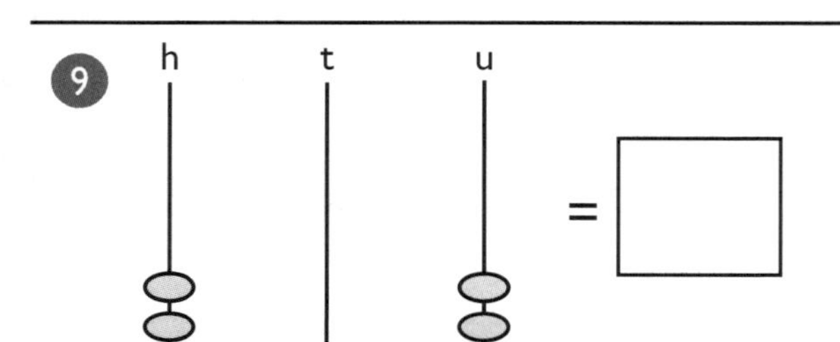

= ☐

10 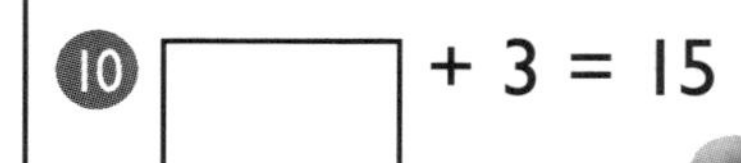 + 3 = 15

Score

Test 35

1. Show 110.

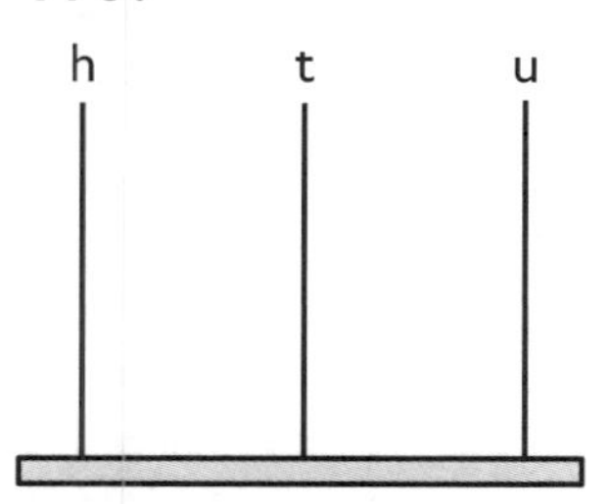

2. Shade three boxes that add up to 7.

1	2	3	4	5

3. 8 < 11 means 8 is less than 11.

Is 20 < 16?

Write yes or no.

4. Draw the other half of this shape.

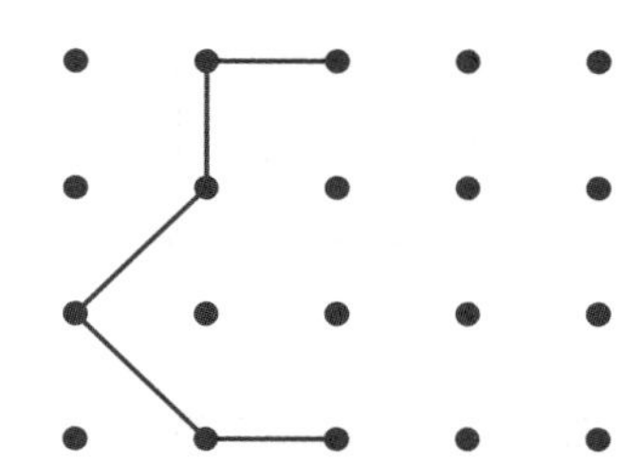

5. Fill in the missing numbers.

30, 28, ☐, ☐, ☐, 20

6. $\frac{1}{4}$ of the box holds 3 crayons.

$\frac{1}{2}$ of the box holds ☐ crayons.

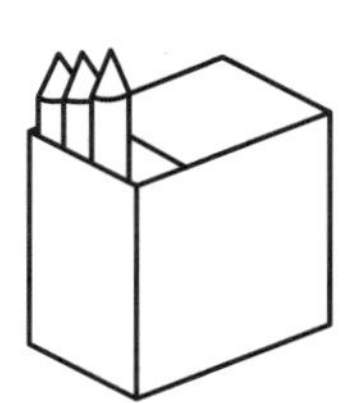

7. 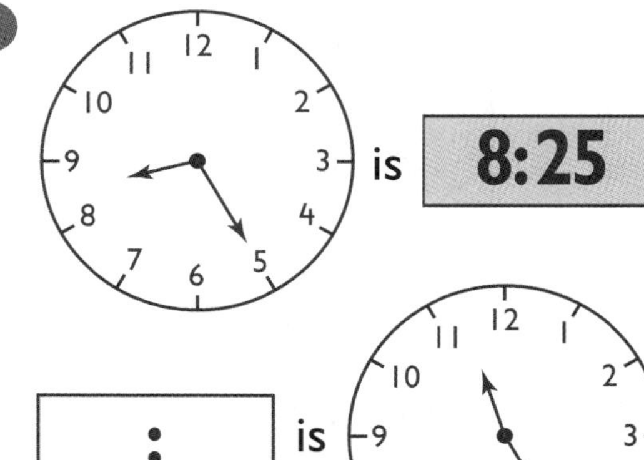

8. 8 + ☐ = 7 + 8

9. 33 − ☐ = 31

10. 24 + ☐ = 30

Score

Test 36

1.  10 more than 28 is 

2. 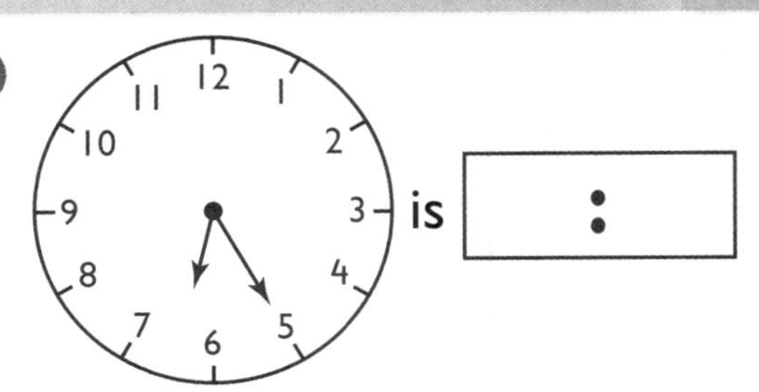

3. Shade three boxes that add up to 8.

1	2	4	4	5

4. Shazia eats $\frac{1}{4}$ of a cake each day. How long will it take her to eat the whole cake?

☐ days

5. Fill in the missing numbers.

33, 31, ☐ , 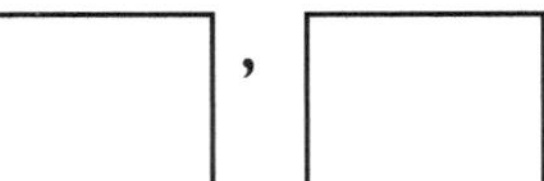 , 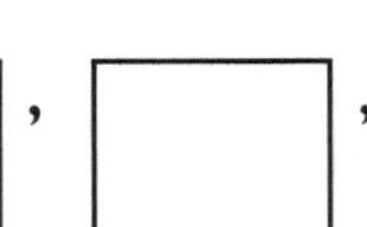 , ☐ , 21

6. Show three hundred and seventy-two.

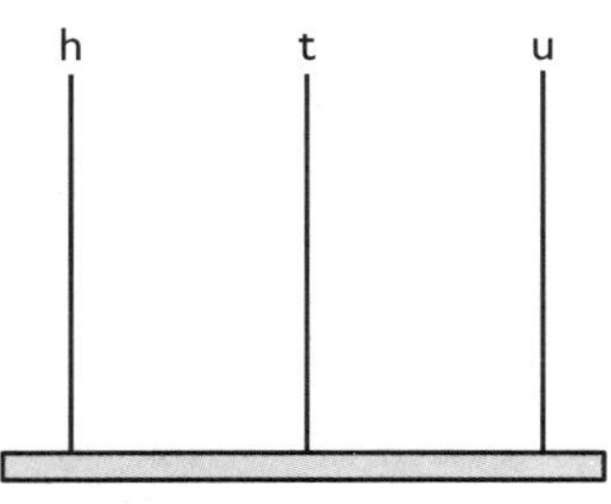

7. In a class there are 30 children. 18 are boys. How many girls are there?

☐ girls

8. 10 less than 38 is 

9. (4 + 9) − 4 = ☐

10. 18 − ☐ = 14

Score

Test 37

1 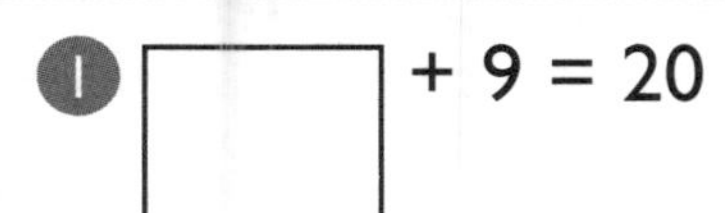 + 9 = 20

2 $\frac{1}{2}$ of 16 = ☐

3 Share 12 sweets equally between 4 children.

Each child has ☐ sweets.

4 In a class there are 17 boys and 20 girls.
How many children are there altogether?

☐ children

5 Draw the missing shapes.

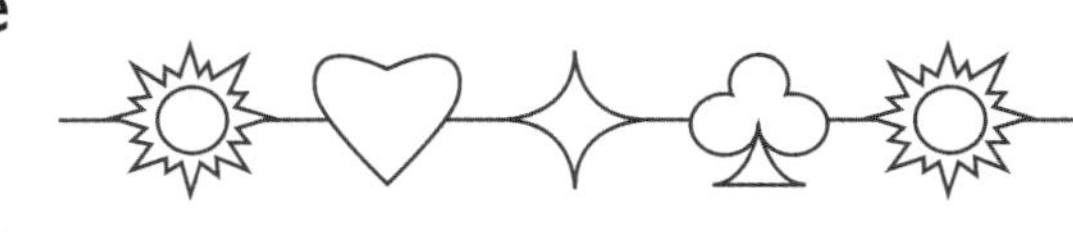

6 Fill in the missing numbers.

7 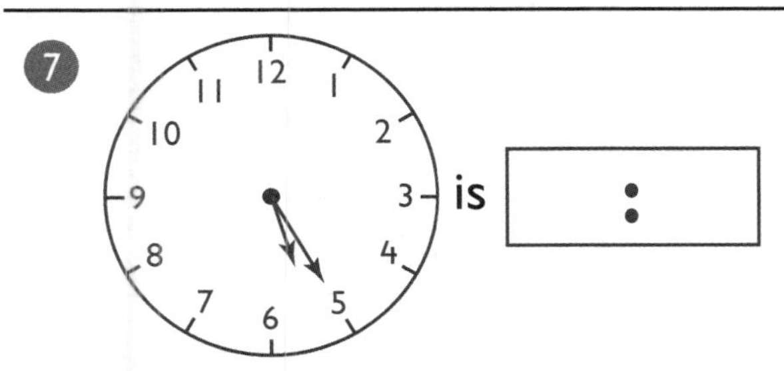

is ☐ : ☐

8 19 – 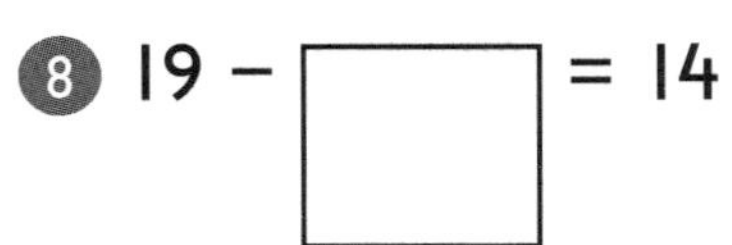

9 Show 402.

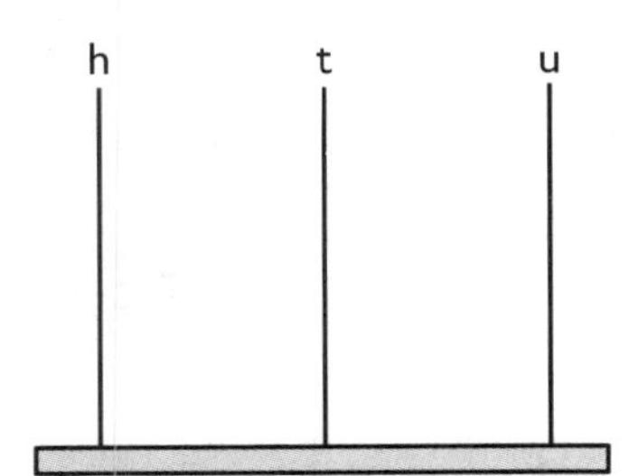

10 24 + ☐ = 26

Score

Test 38

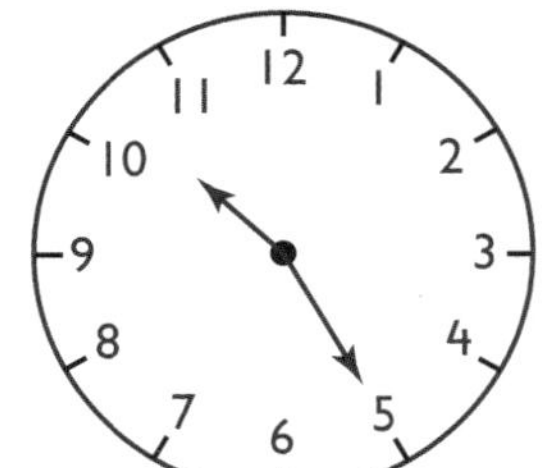

is 25 minutes past [ ]

2. [ ] + 6 = 29

3. The faces of a cuboid are [ ] (flat or curved).

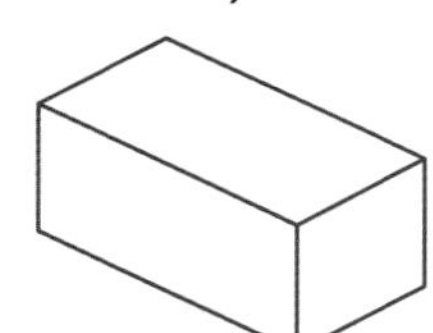

4. Share 20 chocolates equally among 4 children.

Each child gets [ ] chocolates.

5. What number is 1 more than 99? 

6. This shape is a  (square, sphere or cube).

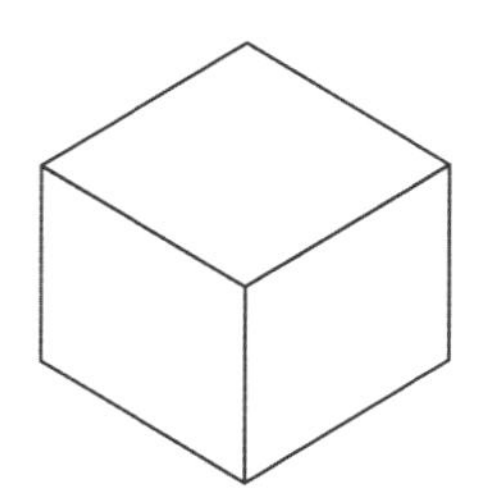

7.

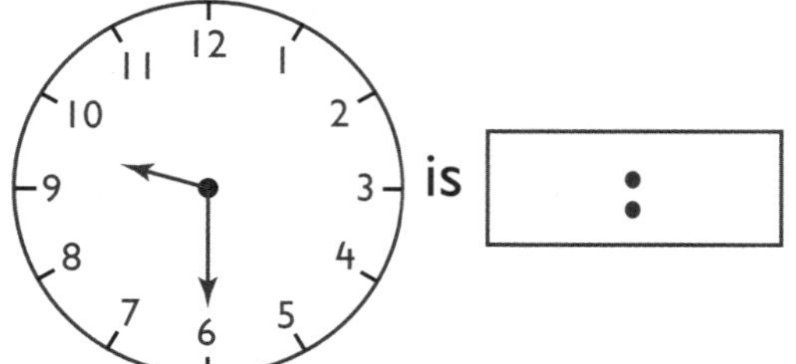

8. Fill in the missing numbers.

40 [ ] [ ]
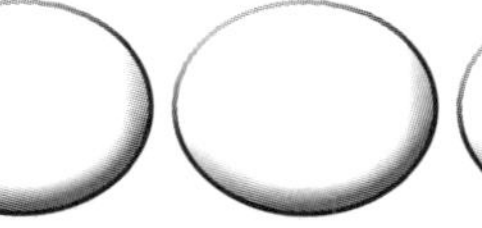

9. 26 − [ ] = 22

10. $\frac{1}{4}$ past 8 is

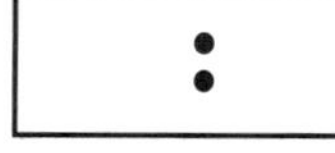

Score

Test 39

1. Show 106.

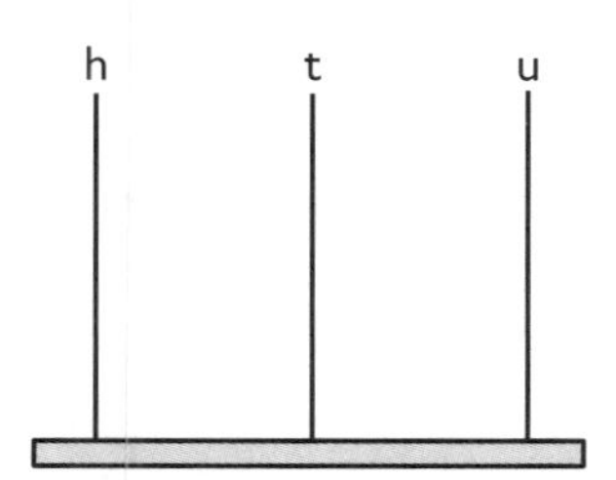

2. 10 more than 67 is 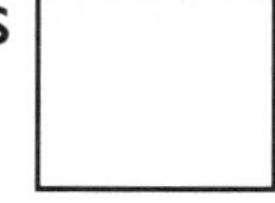

3. 60 minutes = 1 hour

$\frac{1}{2}$ an hour = ☐ minutes

4. Deepa has 30 comics.
Raymond has 16.
How many more has Deepa than Raymond?

 comics

5. $\frac{1}{2}$ past 4 = ☐ **:** ☐

6. 18 − ☐ = 4

7. This shape is a ☐

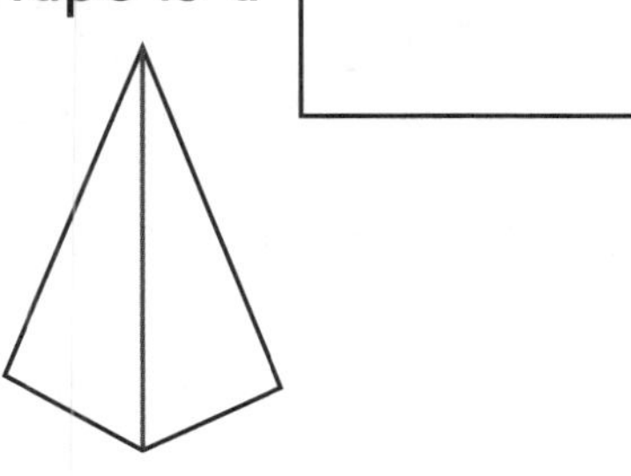

(cuboid, cone or pyramid)

8. Share 12 chocolates equally between Maria, Linda and Colin.

Maria gets ☐ chocolates.

9. Fill in the missing numbers.

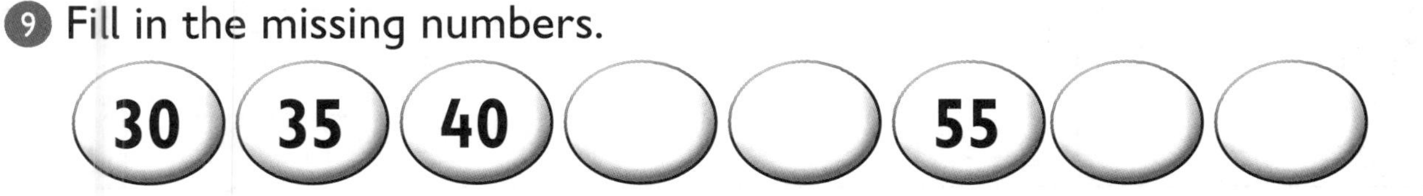

10. Who is lighter, Tom or Colin? 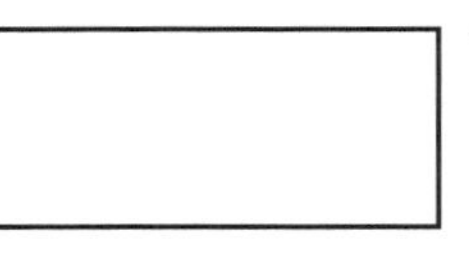

Score

Test 40

1. Shade three boxes that add up to 9.

5	2	3	4	5

2. Draw the other half of this shape.

3. 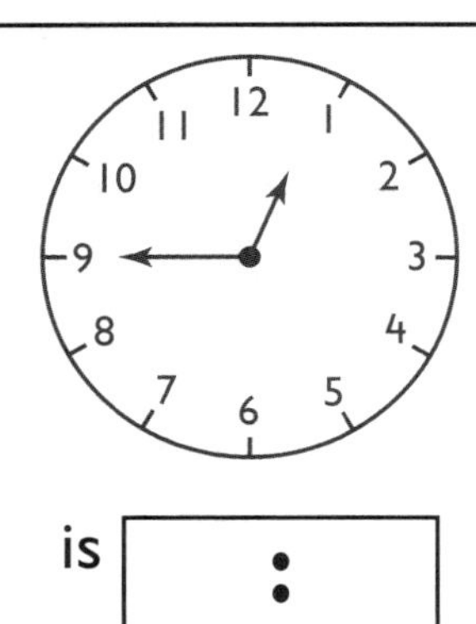

is [:]

4. 5 minutes to 12

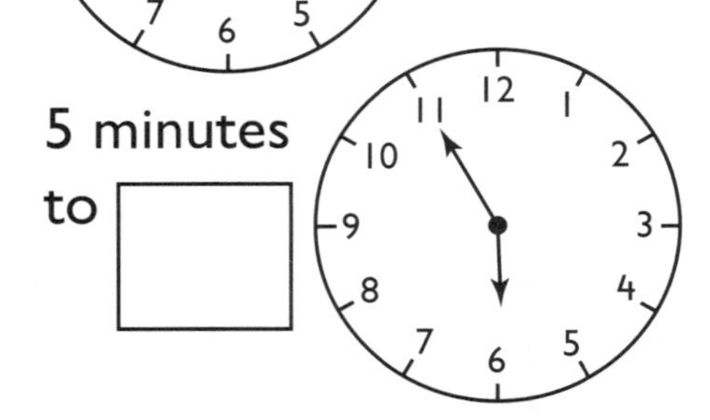

5 minutes to [ ]

5. Fill in the missing numbers.

36, 39, 42, ___, ___, 51, 54, ___, ___, ___

6. Write the number:

one hundred and six

[ ]

7. How many days are in 2 weeks?

[ ] days

8. $(10 - 1) + 3 =$ [ ]

9. $\frac{1}{2}$ past 12 is [:]

10. $18 -$ [ ] $= 1$

Score

Answers

	Test 1	Test 2	Test 3	Test 4	Test 5
1	8		12:00	1	10, 15, 20
2	5	6, 8, 10	30	6	18
3	8	(b)	8	1 2 0 5 3	John
4	(c)	8	20	41	4
5	8	2; 4; 6	14, 16, 18	12, 15, 18	10
6	24	30		(b)	(b)
7	(c)	8:00	5		(a)
8	4; 7	8	4	6	4
9	8	3	4	4	6
10	2	2	(c)	5	30

	Test 6	Test 7	Test 8	Test 9	Test 10
1	8	6	20, 30, 40	4 tens; 0 units	60
2	$\frac{1}{2}$ past 9	Any two blocks should be shaded.	6	11	40
3	(c)	16	(c)	4	5 minutes; 7 o'clock
4	1		5 tens; 3 units	Any three blocks should be shaded.	Any four blocks should be shaded.
5	5; 4; 3	8; 10; 12	15; 18; 19	12	5 minutes; 9 o'clock
6	Gary	Any two pears should be shaded.		1 2 3 4 3	Any four apples should be shaded.
7	4	22, 24, 26, 28	Any three apples should be shaded.		3
8	1 2 0 2 3	10			4
9	50	12	18	5	7
10	7	7	5	23	4

Answers

	Test 11	Test 12	Test 13	Test 14	Test 15
1	2	30	5	11:05	20
2	1 2 0 3 3	1 2 0 1 3	9:05	18, 21, 27, 30	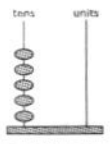
3	5 minutes; 6 o'clock	(b)	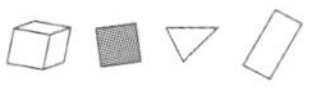	(c)	
4	7	5	13	2	(b)
5	10; 15	7; 6; 4	3 2 0 1 3	12	1 2 0 1 3
6	25	5 past 8	Any six rectangles should be shaded.	A	10
7	Any five blocks should be shaded.	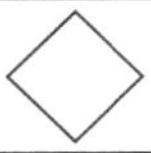	6, 9, 12	Any six circles should be shaded.	9; 12
8	5	4	10:05	23; 22	8
9	50	60	12	70	3
10	4	8	2	10	1

	Test 16	Test 17	Test 18	Test 19	Test 20
1	1 ten; 2 units	4	35	73	16
2	8	4	10 minutes past 11	10 past 4	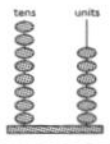
3	Half of the letter should be shaded.	10 minutes; 9 o'clock	B	Half of the letter should be shaded.	(a)
4	(b)	6	(c)	(c)	6
5	11; 13; 15	1 2 0 0 3		(b)	
6	28, 30, 32	6	11; 14	8:10	99, 100
7	14	Half of the letter should be shaded.	3	2	Any one part should be shaded.
8			20; 25; 30	70, 80, 90	15; 18
9	0	1	90	33	12:10
10	3	8	3	26	7

Answers

	Test 21	Test 22	Test 23	Test 24	Test 25
1	96	Yes	2	2	9; 15
2	30, 31, 32	110	100	(c)	90
3	(c)	Any two apples should be shaded.	(c)	Any three toffees should be shaded.	9
4	11; 15	8	Any three parts should be shaded.	5 minutes past 1	
5	8; 6; 4	8	9	7; 11	6:15
6	35		16		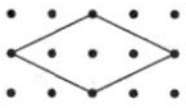
7	1	15 minutes past 10	15 minutes past 9	 1 2 3 4 5	3
8	12	50; 55; 60	0	Yes	13; 15
9		15	sphere	3	4
10	0	15	10	6	5

	Test 26	Test 27	Test 28	Test 29	Test 30
1	1:15	10	8	13	22
2	No	5	10	7	Yes
3	 1 3 3 4 5	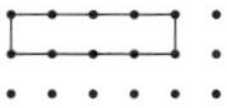	(c)	20 minutes past 10	1:20
4		6	20 minutes; 8 o'clock	Any three parts should be shaded.	10
5	74; 76	27; 30; 39; 42	24; 28; 40; 44	14; 10	50
6	4	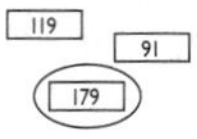 119 91 179	30	32	17
7	14	20	3:20	3	Any three parts should be shaded.
8	124	Any two parts should be shaded.	12	8	0
9	5		15	11	15, 20, 25
10	37	20	12	13	8

Answers

	Test 31	Test 32	Test 33	Test 34	Test 35
1		1	8, 12, 16	7	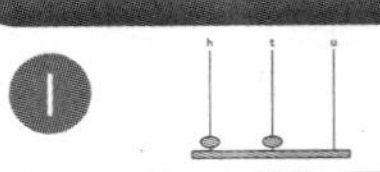
2	12:20	255	16	Any two blocks should be shaded.	1 2 3 4 5
3	3	No	12	214	No
4	18		1 2 4 4 5	1 2 3 4 5	
5	42; 48; 54	12; 16	28; 26	18; 21	26, 24, 22
6	Any six parts should be shaded.	Any three parts should be shaded.	Three-quarters of the rectangle should be shaded.	25 minutes past 3	6
7	1 2 3 4 5	1 2 3 4 5	25 minutes; 10 o'clock	2	11:25
8	15; 10; 5	6; 3; 0		12	7
9	13		303	202	2
10	0	14	6	12	6

	Test 36	Test 37	Test 38	Test 39	Test 40
1	38	11	10		5 2 3 4 5
2	6:25	8	23	77	
3	1 2 4 4 5	3	flat	30	12:45
4	4	37	5	14	6
5	29, 27, 25, 23		100	4:30	45; 48; 57; 60; 63
6		20; 25	cube	14	106
7	12	5:25	9:30	pyramid	14
8	28	5	50; 60	4	12
9	9		4	45; 50; 60; 65	12:30
10	4	2	8:15	Colin	17